T0339409

Energy from Waste

The Power Generation Series

Energy from Waste

Paul Breeze

ACADEMIC PRESS
An imprint of Elsevier

Academic Press is an imprint of Elsevier
125 London Wall, London EC2Y 5AS, United Kingdom
525 B Street, Suite 1800, San Diego, CA 92101-4495, United States
50 Hampshire Street, 5th Floor, Cambridge, MA 02139, United States
The Boulevard, Langford Lane, Kidlington, Oxford OX5 1GB, United Kingdom

Notices
Knowledge and best practice in this field are constantly changing. As new research and experience broaden our
understanding, changes in research methods, professional practices, or medical treatment may become
necessary.

Practitioners and researchers must always rely on their own experience and knowledge in evaluating and using
any information, methods, compounds, or experiments described herein. In using such information or methods
they should be mindful of their own safety and the safety of others, including parties for whom they have a
professional responsibility.

To the fullest extent of the law, neither the Publisher nor the authors, contributors, or editors, assume any
liability for any injury and/or damage to persons or property as a matter of products liability, negligence or
otherwise, or from any use or operation of any methods, products, instructions, or ideas contained in the
material herein.

British Library Cataloguing-in-Publication Data
A catalogue record for this book is available from the British Library

Library of Congress Cataloging-in-Publication Data
A catalog record for this book is available from the Library of Congress

ISBN: 978-0-08-101042-6

For Information on all Academic Press publications
visit our website at https://www.elsevier.com/books-and-journals

  **Working together
to grow libraries in
developing countries**

www.elsevier.com • www.bookaid.org

Publishing Director: Joe Hayton
Acquisition Editor: Raquel Zanol
Editorial Project Manager: Mariana Kuhl
Production Project Manager: Mohana Natarajan
Designer: MPS

Typeset by MPS Limited, Chennai, India

CONTENTS

CHAPTER 1

An Introduction to Energy From Waste

Human activity generates a range of waste materials. Some of these can be reused or recycled, but there are always residues that have no further value in their current form. Often, these residues can be exploited in a combustion plant to generate electrical power, or as a feedstock that is used to make a liquid or gaseous fuel. Where this is possible, it provides both a means of disposal and an additional useful product.

Agriculture produces some of the largest volumes of waste. Many crops, such as cereals, sugar cane or rice leave waste material behind when harvested. Provided it is economical to collect, this type of waste is easily combustible and can be used in power plants designed specifically for the purpose. The raising of animals, too, produces waste. This is not always flammable but can often be converted into biogas using a process called anaerobic digestion. Forestry, like agriculture, produces waste that, if collected, can easily be used to generate electrical power. The paper industry, which relies on forestry, generates several types of waste that have traditionally been used to generate power and heat with the energy used to power the industrial plant producing the paper. Other industries produce specialized wastes, and these can sometimes be converted into energy. Often, however industrial wastes must be disposed of using special techniques, particularly if they are hazardous.

The most ubiquitous and socially most important source of waste, however, is municipal waste, the waste that is produced by households and individuals as they go about their lives. Waste of this type is produced in all societies, but modern advanced societies produce far more than older rural societies. Urban waste, in particular, is produced in massive volumes and its collection and disposal is both costly and time consuming. Some of this waste, such as paper, glass, and metal cans, can be recycled. This involves sorting the waste either before or after collection. Organic residues can be allowed to decompose naturally

Energy from Waste. DOI: http://dx.doi.org/10.1016/B978-0-08-101042-6.00001-7

and then be returned to the soil to provide nutrients. However, there will always remain a significant residue. Exploiting this to generate energy offers a cost-effective and convenient method of disposal.

In the past, the combustion of residual waste, often without electricity or heat generation, has been used as a means of reducing the volume of waste for disposal. The residual ash is then buried in a landfill site. Such processes are wasteful of energy and today this will not be considered appropriate in many jurisdictions and may contravene local regulations. Across the European Union (EU), for example, there are strict rules about how waste must be treated and combustion without energy generation would be considered one of the least desirable options.

Combustion technologies that are capable of utilizing the energy released from the waste to generate electricity and heat offer a much more environment-friendly solution. However today, there are additional considerations to take into account related to atmospheric emissions and global warming. Some waste may be biological in origin; wood, paper, and agricultural products might fit into this category. These can be considered to be renewable. Their combustion, while producing carbon dioxide, does not add to the atmospheric load because they are part of a short biological cycle in which more, similar material will soon be grown and this will reabsorb carbon dioxide from the atmosphere. On the other hand, plastics are often made from fossil-fuel-based materials and so when these are burnt in a combustion plant they add to the atmospheric load of carbon dioxide. This must be taken into consideration when assessing energy from waste projects in regions such as the EU.

There are several ways of generating energy from waste. The simplest and most widely used is to burn the combustible material in a combustion boiler, generating heat that is used to raise steam and drive a steam turbine generator. Power generation depends on the quality of the waste and (its energy content, or calorific value), but the efficiency is generally relatively low at around 25% to 30%. It is possible to raise the efficiency if heat from the plant can be used as well as electricity. This depends on being able to site a power from waste plant close to users of heat, which is not always possible, but where heat can be used, efficiency can be increased to over 40%.

Two more advanced techniques for dealing with waste are pyrolysis and gasification. Both are high temperature techniques that can ensure destruction of complex; sometimes, hazardous organic molecules within the waste. Pyrolysis can produce gaseous or solid byproducts that can be burnt to generate energy or used in other processes. Gasification normally produces a low energy content gas which can also be burnt to generate energy. The gas can be used either in a conventional steam generating boiler or may be used as fuel for a piston engine or gas turbine.

Whatever the process, the generation of power-from-waste is a very specialized industry. The plants must include extensive environmental controls to ensure that they do not release any toxic materials into the environment and are consequently expensive to build. The cost of electricity from a power to waste plant will be high when calculated using the normal economic model for the cost of electricity. Against this the economics of a waste to energy plant does not normally depend entirely on the value of the electricity it produces; plant operators are generally paid for the waste that they process and this helps balance plant economics. In a sense electricity and heat are valuable byproducts and disposal of the waste is the main object. Increasingly, however, waste is being considered as a useful energy resource that should, where possible, be utilized in the most energy-efficient way possible. This is particularly relevant to the advanced economies that produce the most waste.

WASTE TO ENERGY: A HISTORICAL PERSPECTIVE

The combustion of waste probably has a long, unrecorded history but municipal combustion or incineration of waste can be traced back to 1874 when the first incinerator, known as a destructor, was built in Nottingham, United Kingdom, by a company called Manlove, Alliott and Co. The earliest US incinerator was constructed 1 year later, in 1885, when a unit was built on Governors Island, New York. A further 200 were built across the United States in the next two decades, although by 1902 half of these had been abandoned or dismantled. It is likely that similar units were in use in many European countries by, or soon after the turn of the 20th century. Meanwhile in the United States as the 20th century advanced and the problem of waste disposal in cities became more onerous, it became common for apartment buildings to have their own incinerators.

Energy recovery with incineration appears to have soon followed the original destructor-style incinerators. The first plants of this type were built in the early 20th century. The earliest recorded installations include a Swedish energy from waste plant which opened in 1904. Similar plants are known to have operated elsewhere in Europe and in the United States in the early years of that century.

The use of combustion and waste to energy plants was concentrated in the developed countries that produced the most waste, and particularly those with large urban populations where the problem of waste management were the most acute. Europe has the strongest tradition of waste to energy technology, illustrated by the fact that at the end of the 20th century two European companies, Martin GmbH based in Munich and the Swiss company Von Roll (now Hitachi Zosen Inova) accounted for close to 70% of the market for mass-burn waste to energy plants, the dominant technology, then and today. Japan adopted waste to energy technology early in the 20th century too and the United States has a long, uneven history of energy generation from waste.

In Europe, there is also a long history of the use of district heating in urban areas and many waste plants were designed to produce heat or heat and power. Elsewhere district heating is less popular so that waste to energy plants are more often designed to produce electricity from steam.

Early waste to energy plants were often relatively polluting by today's standards. Overall emission standards were tightened in many countries in the middle of the 20th century and later in the century there was additional regulation to control specific toxic emissions such as dioxins and heavy metals. These represent a particular danger in waste incineration plants where the content of the fuel composed of municipal or industrial waste—or sometimes a mixture of both—may be diverse. As a consequence, modern waste to energy plants have complex emission control systems designed to prevent toxic emissions being released into the atmosphere.

Any waste to energy plant requires a regular supply of fuel if it is to operate successfully. For this reason, the history of energy from waste is also closely linked to the organized collection and disposal of urban waste. This began at the end of the 19th century. Although combustion was one option that was adopted in some countries, it was not widespread and instead the waste was often consigned to dumps. In the United States, this meant off-loading the waste into wetlands with a mixture of ash and soil. The concept of the landfill waste site had been born.

As the problem of waste disposal grew, so conventional landfill became the favored method of disposal. The first identifiable modern landfill site in the United States began operating in 1935 in California. Since then, over much of the last century, this became the favored method of waste disposal in many countries. Landfill waste usually contains a large percentage of organic material, and in a landfill, this undergoes anaerobic fermentation which produces methane as a byproduct. Carbon dioxide is also generated. The methane can be dangerous, leading to fires on landfill sites; by the end of the last century, it was also recognized as a potent greenhouse gas. Legislation has followed, and in many countries, the methane that is produced in landfill sites must be captured to prevent it leaking into the atmosphere. This gas can be used as fuel, and landfill sites often have gas turbines or reciprocating engines that use the gas to generate electricity.

The use of agricultural wastes for power generation is relatively recent, driven in part by their renewable credentials. In the past, agricultural waste was often burnt; straw burning was common in British fields until towards the end of the 20th century. Since then, the waste has often been collected for reuse. However, the wood and paper industries have a long history of using their wastes for energy production. There is also a tradition of using animal wastes to produce a biogas for energy production. However, this is usually localized and at a small scale.

The level of exploitation of waste-to-energy plants varies from country to country. They have been used widely in parts of Europe, where waste has been burned since the end of the 19th century, and they form a major part of Japan's waste disposal strategy. In contrast, the United Kingdom and the United States have only adopted the technology patchily, while some nations, such as Slovenia, Romania, Bulgaria and Greece, do not incinerate any waste. This is partly tradition, but environmental concerns about the emissions from the plants have resulted in local resistance to their construction in some parts of the world. More advanced emission treatment processes may make it easier to build these plants in the future.

Some countries have used the incineration of waste extensively. In Sweden, for example, 14% of municipal waste is put into landfill sites, 41% recycled and 45% is incinerated. By contrast, the United Kingdom disposes of 49% of its waste in landfill sites, recycles 39% and incinerates 12%. The Netherlands, meanwhile, incinerates 33%, recycles 64% and only disposes of 3% in landfill sites. Within the EU Denmark has the highest incineration rate, 54%. Meanwhile in Germany 62% of waste is recycled and 38% incinerated. In Japan, around 60% of solid waste is incinerated.

Where they are employed, power-from-waste plants generally burn domestic and urban refuse—called in this context municipal solid waste (MSW)—using the resulting heat to generate steam to drive a conventional steam turbine. Some heat may also be used for district heating. MSW can also be sorted and treated to produce a compacted fuel called refuse-derived fuel (RDF) which can be burned in a power station. A further category, solid recovered fuel (SRF) is like RDF but with specific properties defined such as its calorific value, moisture content and particle size. SRF can be used in some conventional power plants. Some industrial waste may be treated in the same way as domestic waste. However, industrial wastes are likely to contain toxic materials which have to be handled using special procedures. Where such care is not required, they can be dealt with in the same way as urban waste.

WASTE AS A RENEWABLE RESOURCE

As the issue of global warming has risen up the international agenda, so the need to identify and exploit renewable and sustainable sources of nonfossil fuel energy has become more important. Although the main emphasis of this is on technologies such as solar and wind energy, any potential renewable energy source is of interest today. Municipal waste contains a large percentage of material that is biological in origin. Paper, cardboard packaging, garden waste, and food residues as well as wood from building demolition or other sources is all biogenic and as such may be considered renewable.

The renewable content of waste varies. Since waste is a mixture of materials, some of which are derived from fossil fuels, it is important to determine what fraction of waste is biogenic in origin and therefore might be considered renewable. This will vary from place to place and will depend on the time of the year because waste varies with the seasons. A greater degree of consistency can be obtained if waste is sorted so that much of the nonbiogenic material is removed and mainly renewable material remains. Depending on local regulations, waste that is 90% biogenic may be considered similar to biomass from a renewable fuel perspective.

Since the Paris climate change agreement of December 2015, under which all countries will set out their plans for tackling carbon emissions, most governments are interested in recruiting any potential renewable resource to help meet their emissions targets. The renewable portion of waste fits into this category and this is creating an additional interest in energy from waste. Waste has the added bonus that it can provide a continuous supply of energy, not intermittent like wind or solar energy. On the other hand, the amount of energy available is small and is likely to get smaller as societies demand that more and more waste is recycled.

GLOBAL PRODUCTION OF POWER FROM WASTE

According to figures collated by the World Energy Council, based on data from the International Renewable Energy Agency, the total global installed generating capacity for waste to energy plants in 2015 was 12,912 MW, while total electricity production in 2014 was 40,131 GW h. These figures are shown in Table 1.1 which also shows a breakdown by country for all those for which figures are available. Data for waste to energy is sparse in many parts of the world and so these figures should be considered as approximate.

From the figures in the table, the most notable countries in terms of waste to energy plant capacity are the United States with 2254 MW, Germany with 1888 MW, and Japan with 1501 MW. Other countries with significant capacities include France with 872 MW, Italy with 826 MW, the United Kingdom with 781 MW, the Netherlands with 649 MW, Chinese Taipei with 629 MW, and Sweden with 459 MW. Most European countries have some waste to energy plants; Ireland has one as does Poland and Hungary. Meanwhile, France has 126, Germany 99, Italy 43, Sweden 33, and Switzerland 30. Other notable entries in the table are the Island states of Chinese Taipei and

Table 1.1 Global Generating Capacity and Production of Waste to Energy Plants by Country

Country	Waste to Energy Generating Capacity in 2014 (MW)	Waste to Energy Electricity Production in 2014 (GW h)
Austria	539	285
Azerbaijan	37	174
Belgium	247	810
Canada	34	89
Chinese Taipei	629	1596
Czech Republic	45	88
Denmark	325	885
Estonia	210	–
Finland	–	441
France	872	1824
Germany	1888	6069
Hungary	22	137
India	274	1090
Indonesia	7	32
Ireland	17	68
Israel	6	14
Italy	826	2370
Japan	1501	6574
South Korea	184	564
Lithuania	10	29
Luxembourg	17	34
Malaysia	16	17
Martinique	4	23
Netherlands	649	1909
Norway	77	176
Portugal	77	240
Qatar	25	110
Singapore	128	963
Slovenia	11	22
Spain	251	686
Sweden	459	1626
Switzerland	398	1102
Thailand	75	201
United Kingdom	781	1422

(Continued)

Table 1.1 (Continued)		
Country	Waste to Energy Generating Capacity in 2014 (MW)	Waste to Energy Electricity Production in 2014 (GW h)
United States	2254	8461
Uruguay	1	–
World	12,912	40,131
Source: World Energy Council/IRENA.[1]		

Singapore; both have limited space for landfill and use combustion extensively to control and manage waste.

The production of electricity from waste across the globe broadly follows the pattern of installed capacities. The United States produced 8461 GW h in 2014 from its waste to energy plants, while Germany produced 1888 GW h and Japan 6574 GW h. It is notable that the electricity production in Germany is lower per unit of capacity than in either the United States or Japan. This likely reflects the fact that many plants in Germany are combined heat and power plants whereas those in Japan and the United States are primarily electricity generating plants. Where heat is being used the electricity output will be lower than if all the heat is being used for power generation.

Although Europe has the most highly developed waste to energy infrastructure, interest and investment is growing in the Asia-Pacific region. Japan is the most highly developed nation in the region in terms of waste to energy use but the use of the technology in China is growing rapidly. Although no figures for China are included in Table 1.1, an alternative source indicates that in 2015 there were 20 operating waste to energy plants in 15 cities in China, the largest producing 136 GW h of electrical power each year.[2]

The total amount of waste being generated in 2012 was has been estimated to be 3.5 m tonnes per day.[3] By 2025 this is expected to reach over 6.1 m tonnes each day. Global peak waste production is not predicted to be reached until the end of the century.

[1]World Energy Council, World Energy Resources: Waste to Energy 2016. The figures in the report are derived from the International Renewable Energy Agency 2016 statistics.
[2]Chinese Waste-to Energy Market Experiences Rapid Growth During Last Five Years, Liu Yuanyuan, Renewable Energy World, April 28, 2015.
[3]World Energy Council, World Energy Resources: Waste to Energy 2016.

CHAPTER 2

The Politics of Waste

As the human population has grown and as the people of the world have congregated more and more in towns and cities, so the amount of waste generated by the world's human population has grown. The move from the country to towns has been common as nations have industrialized over the past two centuries, and it is continuing today. In 2014, the proportion of the world's population that lived in cities was 54% according to the UN's *World Urbanization Prospects*, and this is expected to rise to 66% by 2050. Of course, waste is generated by both rural and urban populations, but it is the waste generated in cities and towns that causes the most pressing issues.

For the past century and more, the waste generated in cites has been seen as a problem for which the optimum solution was easy and rapid disposal. In many cases, this has meant placing waste in rural-waste dumps which are often, but not always, landfill sites. (There are still mountains of waste around some of the world's cities although these are often called landfills.) The other alternative in advanced economies, particularly those that have limited space for landfill, is to burn the waste.

In the 21st century, the approach to waste is changing and at the same time waste has increasingly become a political issue. As the world's population grows, and as resources become strained, and in many cases depleted, waste is increasingly being seen as a resource that must be exploited as efficiently as possible. This has led to the evolution of strategies such as zero waste and the circular economy. What these encourage is the reuse of all waste. Taken to its limit, this would leave no waste to be converted into energy and the concept of waste to energy would become redundant.

A strategy such as zero waste involves changing the approach to manufacturing that has been in use ever since the industrial revolution. The traditional system—in this context often called the linear system—involves resource extraction, manufacturing, distribution, use, and then

Energy from Waste. DOI: http://dx.doi.org/10.1016/B978-0-08-101042-6.00002-9

disposal. In a zero waste approach, the final stage, disposal, is no longer permitted. Instead, everything must be reused in some way. This, in turn, changes the way products are manufactured. The new chain is called the circular economy because when a product has reached the end of its useful life, it is returned to the beginning of the production chain through some form of recovery. This might be remanufacturing of the original product; it might involve recycling of the components; for organic waste such as food waste it would probably mean composting. There might still be a residual amount of waste but the quantity would be small.

Zero waste is a utopian ideal. In practice, it is unlikely ever to be achieved. Nevertheless, its goals are important and they are affecting the way in which waste is managed in many parts of the world. This has led to governments adopting waste management policies that are based on aspects of the zero waste. In some cases, these are aspirational; in others, they are enshrined in legislation. This is likely to have an important influence on the management of waste in the future.

WASTE MANAGEMENT HIERARCHY

The Waste Hierarchy is a framework for waste management that has become widely adopted in recent years, at least in some regions. This is especially true of the European Union (EU) where it is part of the EU Waste Framework Directive. Within the framework, the hierarchy sets out, in order of priority, the way in which waste should be handled for to achieve the minimum environmental impact.

The principle behind the waste hierarchy is that any material that is consigned to waste should be used in the best, most efficient and most environmentally benign way possible. With this in mind the hierarchy sets out five (or sometimes six) levels of waste management. The hierarchy is shown schematically in Fig. 2.1. In descending order of preference, these five levels are as follows:

Prevention: The idea behind prevention, the first stage in the waste management hierarchy, is that products are designed from the outset to generate less waste. This may be by designing so that a whole product or components within it are reusable. Extending the lifetime of products will also reduce the amount of waste generated. For products that are consumed, it will address issues such as

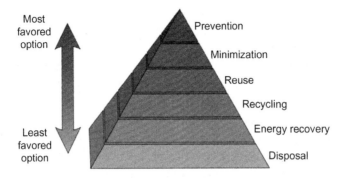

Figure 2.1 The waste management hierarchy. Source: Wikipedia Commons.

packaging which should be minimized, or avoided completely where possible. *Minimization* addresses the same issues but is sometimes considered as a separate stage in the hierarchy although it broadly overlaps with prevention.

Reuse: This is the second stage in the hierarchy. When a product reaches the end of its life, it should wherever possible be reused. Simple examples of this include the reuse of bottles after their contents have been consumed. It has been traditional, in the United Kingdom, for example, to provide doorstep delivery of milk in glass bottles that are collected and reused by the milk delivery service. This can be extended to much more complex products through a process called remanufacturing. This seeks to rebuild a product to match its original specifications using a combination of original, recycled, and new parts. Refurbishment and repair of products also fit into the reuse category.

Recycling: When there is no longer the possibility of reusing a product or component, the next preferred option is to recycle it. This will involve using the material in the product to make something new. Obvious recycling candidates include glass, metals, and paper, all of which can be exploited to create new glass, metal, or paper products. Complex products that contain many components such as vehicles, household white goods or electronic and electrical devices should be dismantled, and the best use sought for each of the dismantled parts. Recycling is also applied to the organic component of waste. Material such as food waste and other similar household of commercial waste should be allowed to decay into compost so that the nutrients that remain within the residual material can be returned to the soil.

Other (including energy) recovery: Once all reuse and recycling options have been exhausted, then energy recovery can be considered. This should be the most efficient system compatible with the waste category available. For much organic waste some form or anaerobic digestion to produce a biogas may be the best option. Gasification and pyrolysis are also considered to be relatively energy efficient. If the waste is to be burnt, then using it in a combined heat and power facility is preferred. If, finally, none of these options are viable then combustion with energy recovery for electricity generation alone may be considered.

Disposal: The terminal option is disposal of the waste. This is considered as a last resort. Disposal might involve incineration of the waste without any energy recovery or sending it directly to a landfill site to be buried. Often the second will follow the first, with landfill taking care of the ash residue from incineration.

CIRCULAR ECONOMY/ZERO WASTE

Although the waste management hierarchy seeks to make optimum use of the waste that is produced today, zero waste strategies and the circular economy are ambitious concepts for production and manufacturing in the future. The two terms are essentially the two sides of a single coin since each implies the other.

As already noted, both are ideals that are unlikely to be achieved in practice, at least not in the foreseeable future. However, the adoption at a government level of the concept of zero waste or of a circular economy has wide ranging implications for industry. For example, it implies that all durable products should be designed with as long a lifetime as is feasible and constructed using components and techniques that enable future remanufacturing. Similarly, for consumable products such as foods or a variety of household goods, it implies the use of packaging materials that can be reused or at least easily recycled in a way that enables the material of the packaging to be remanufactured.

For the waste to energy industry, this clearly implies a severe reduction in the amount of waste material that would be available for conversion into energy. However there are probably limits to the amount that even the best zero waste economy can achieve. As an example, the Scottish Government produced a zero waste plan in 2010 which targeted 70% recycling by 2025 and less than 5% of waste material going to landfill. The EU has also adopted a circular economy package that targets the recycling of 65% of municipal waste by 2030 and at the same time the recycling of 75% of packaging waste. The use of landfill is also restricted. Other countries and regions have more onerous targets. For example, New Zealand is hoping to achieve a zero waste target by 2020 of no landfill and no waste incineration.

Most of these targets should be achievable: Although the EU as a whole only recycled 35% of its municipal waste in 2010, according to European Environment Agency, Austria managed 63%, Germany 62% and Belgium 58%. Even at the target levels, however, there will still be a significant quantity of material that cannot be recycled. Much of this is likely to be suitable for energy recovery, particularly if landfill is prohibited for most types of waste. It might also turn out that this residual waste is more predictable in its composition, making the task of energy recovery easier.

ENERGY FROM WASTE STRATEGIES IN A CIRCULAR ECONOMY

Where governments adopt zero waste or circular economy targets, the question also arises as to the best means of recovering energy from the residual material, and indeed if this is an efficient environmental strategy. Is it better from an environmental perspective to recover energy or to simply bury the waste?

The first level of differentiation is between competing methods of energy recovery. It is generally considered better to convert waste into a new fuel, be that gaseous or liquid, than to burn it. Methods to generate a new fuel include anaerobic digestion and various ways of treating waste to generate liquid fuels. Both pyrolysis and gasification can also produce fuel products, and these may be considered preferable to the most common solution, combustion. However, the emissions implications of these processes must also be taken into account.

Combustion, sometimes called mass burn, is generally relatively inefficient because plants are small by power plant standards and because the fuel does not contain anywhere near the amount of energy of conventional fuels. However waste to energy combined heat and power stations, such as are common in some parts of Europe, are more efficient than stations that can only utilize the heat for power generation. This latter option is therefore the least best option for energy recovery. Today, it is also the most common.

How does the combustion of waste to generate energy compare environmentally with burying the same waste in a landfill site? If one considers a typical bag of household waste, this will contain a mixture of biogenic waste and waste materials such as plastics that were produced from fossil fuels. If all this is burned, both the biogenic waste and the plastics will produce carbon dioxide that will be released into the atmosphere by the waste to energy plant. The carbon dioxide produced by the biogenic waste can be ignored on the basis that biogenic waste is a form of biomass and is renewable. However the carbon dioxide from the fossil-fuel-based waste cannot be ignored and must be considered to add to the atmospheric load.

If the same waste is buried in a landfill site, the fossil fuel originated waste will be sequestered since it is unlikely to break down naturally once buried. However, the biogenic waste will breakdown in the conditions of the landfill. Some of this will lead to carbon dioxide production, but as before this can be ignored since it is essentially renewable. However some will lead to the production of methane. In most cases, the larger part of this methane will be captured and can be used to generate energy. However, crucially, not all of it will be captured. Methane is a much more potent greenhouse gas than carbon dioxide and its leakage into the atmosphere of methane is expected to tip the balance. It is likely, therefore, that there will be less greenhouse gas impact from burning the waste than from relegating it to a landfill site.

WASTE EXPORT

Another political issue is that of waste export. Under EU regulations, for example, it is not permitted for a country to export unsorted municipal waste. However waste can be exported if it has undergone some form of sorting. The most usual form of waste for export is refuse-derived fuel or solid recovered fuel. Again under EU rules, waste of this sort can be exported to another Organization for Economic Cooperation and Development (OECD) country for energy recovery. However no category of waste can be exported for disposal.

In spite of these seemingly strict rules, the EU has exported large quantities of waste such as plastics and paper to countries such as China. Meanwhile, electronic waste is exported to Africa. In terms of overall volumes, the quantities of waste traded in this way are relatively small. Even so they are politically divisive. In addition, the export of wastes in this way means that the country of origin loses any benefit that might be derived from the waste, be that the production of energy or the recycling into reusable material. For some countries this is already an issue, and it is likely to become more important in the future. Waste exporting is normally considered out of bounds under zero waste strategies.

All these considerations mean that, as noted at the outset of this chapter, waste has become a political issue. Although this should ultimately lead to better use of waste and a reduction in the quantity produced, it does make the navigation of waste by companies that wish to become involved in its disposal more complicated than in the past.

CHAPTER 3

Waste as a Resource

Human societies—and the world's nations—produce waste in varying quantities depending upon their level of technical advancement and their standards of living. The most advanced, richest societies produce the most waste, while less developed countries produce lower volumes. For all societies, however, waste represents a problem in need of solutions. Energy from waste offers one potential solution.

Setting aside the political issues examined in Chapter 2, waste management and energy from waste can be complementary to one another. Generating energy from waste offers one of the most attractive means of tackling waste management which, while it cannot provide a complete solution, can offer a way of handling a significant proportion of the waste generated. Ironically, the prospects are often better in nations with higher standards of living because the quality of the waste produced is higher.

Waste quality varies considerably with income level of the society, region, or household producing it. For example, low income societies often rely on markets rather than supermarkets for their produce and less of what is bought is prepackaged. These societies tend to generate more organic waste which has a relatively low energy content. Members of high income societies often buy produce and products packaged, and can afford more consumer goods. This leads to high levels of packaging waste and relatively less organic waste. The waste is both easier to recycle and, when it is not being recycled, contains more energy than the waste from the lower income society.

Waste also varies within a particular society depending upon its source. Most of the waste that will reach a waste to energy plant is classified as municipal solid waste (MSW). This usually means waste collected by a refuse collection system and disposed of through a waste management system—although definitions can vary. However there are other important sources of waste including process waste, medical waste and agricultural waste. Not all of these are useful for

Energy from Waste. DOI: http://dx.doi.org/10.1016/B978-0-08-101042-6.00003-0

energy production and these wastes may be handled outside a local or regional waste management system.

WASTE CLASSIFICATIONS

The primary source of waste for waste to energy plants is MSW. What this actually means can vary from country to country. According to the European Union, MSW is waste from households, as well as other waste which, because of its nature or composition, is similar to waste from households. Another common definition is waste from households and commercial organizations generated in a given municipal area. The waste generated by offices and by public institutions will generally fall into this category, even if it is not specifically included by definition. In addition, there are some other categories that may be classed as MSW. The most important of these are industrial waste and waste from construction and demolition sites. Municipal services will also generate waste that falls into the MSW category if only by virtue of being collected alongside the household and other MSW. The precise composition of MSW is important because it will affect its energy content and this will influence its value as an energy source.

The World Energy Council (WEC)[1] identifies five categories of MSW:

Residential: this category includes household food waste, paper, cardboard, plastics, textiles, leather, garden waste, wood, glass, metals, ash, and large items such as consumer white goods, electronic devices, car tires, and batteries.
Industrial: housekeeping wastes, packaging, food wastes, wood, steel, concrete, bricks, ashes, and hazardous wastes.
Commercial and institutional: paper, cardboard, plastics, wood, food wastes, glass, metals, special wastes, hazardous wastes
Construction and demolition: wood, steel, concrete, soil, bricks, tiles, glass, plastics, insulation material, hazardous waste
Municipal services: street cleaning, landscape gardening and tree waste, sludge, waste from recreational areas.

Some of the materials under each category are of no use in an energy from waste plant, while others are. Segregation or sorting is therefore important if the energy content of waste is to be controlled.

[1]World Energy Council, World Energy Resources: Waste to Energy 2016.

MSW from all these sources can be classified as organic and inorganic. A more detailed breakdown is given by the World Bank. The Bank's classification includes six categories, organic, paper, plastic, glass, metals, and other. The global breakdown of waste according to these categories is shown in Table 3.1.

Globally, the most common type of waste is organic waste which accounts for 46% of the total. This includes food waste, garden waste, wood, and process residues. Paper waste is, strictly, organic too, but for the purposes of MSW classification, is it more useful to identify it separately because it has a much higher energy content than most organic waste. Under paper waste, the World Bank includes paper, cardboard, newspapers, books, magazines, paper cups, and any shredded paper. Paper waste makes up 17% of global waste. Plastic waste, which includes plastic bottles, packaging, containers, bags, lids, and cups, contributes 10% of global waste. All this is primarily derived from fossil fuel. Glass, primarily bottles but also including light bulbs and glassware accounts for 5%. If separated from other waste this can be recycled. Metal waste, 4% of the total, includes cans and tins, foil, old appliances, and a range of other consumer goods and building materials. Again this part of any waste stream should normally be separated and recycled. The final category, other, which is the second largest after organic waste at 18% of the total, covers everything else. Typical ingredients are textiles, leather, rubber, laminates, ash, and a range of inert materials. Some of these are suitable for generating energy, others require alternative treatment.

In addition to MSW, there are other types of waste that can be of value as energy sources. These include process wastes which might be chemical wastes or material left over from a manufacturing process, medical waste, and agricultural waste. Of these, the most important is

Table 3.1 Composition of Global Municipal Solid Waste (MSW)	
Material	Proportion of Municipal Solid Waste (%)
Organic	46
Other	18
Paper	17
Plastic	10
Glass	5
Metal	4
Source: World Bank.	

agricultural waste. Most agricultural waste comprises the residue after crops that have been harvested, and this type of material can easily be converted into energy.

Depending upon the location, another category that may need to be considered separately is construction and demolition waste. The will comprise building rubble, masonry, and concrete as well as some wood and metal. In some cities, this can account for around 40% of the total waste stream according to the World Bank. The size of this portion needs to be identified when planning waste management strategies.

REGIONAL VARIATIONS IN MUNICIPAL SOLID WASTE

The figures in Table 3.1 show the average global composition of MSW, but this varies enormously from region to region and country to country. It is affected by a range of factors including the level of economic development and affluence of the inhabitants, the culture, and the climate. As a general guideline, low income countries tend to produce proportionally more organic waste than richer countries. For the latter, meanwhile, the amount of paper, packaging, and plastic in the waste increases compared to the organic material. This increases the overall volume of the waste. The absolute quantity of waste increases with affluence too. Richer nations produce more waste than the less rich.

Table 3.2 contains figures from the WEC's 2016 survey of world energy resources that show the amount of waste generated in different

Table 3.2 Regional Daily Production of Waste		
Region	Total Urban Municipal Waste Generation in 2012 (t/day)	Total Urban Municipal Waste Generation in 2025 (t/day)
Africa	169,120	441,840
East Asia and Pacific	738,959	1,865,380
Eastern and Central Asia	254,389	354,811
Latin America and the Caribbean	437,545	728,392
Middle East and North Africa	173,545	369,320
OECD	1,566,286	1,742,417
South Asia	192,411	567,545
Total	3,532,255	6,069,705
Source: World Energy Council.[2]		

[2]World Energy Council, World Energy Resources: Waste to Energy 2016.

regions in 2012. In absolute terms, the countries of the OECD produced the most waste, 1,566,286 t each day or around 44% of the global total. East Asia and the Pacific region generated the second largest amount of daily waste at 738,959 t/day, followed by Latin America and the Caribbean with 437,389 t/day. South Asia generated 192,411 t/day, the Middle East and North Africa 173,545 t/day, and Africa 169,120 t/day, the smallest daily amount. Total global daily waste production in 2012 was estimated to be 3,532,255 t/day.

These figures do not tell the whole story, however, because waste production will depend on the size of the population and so the production per capita will tell more about the level of development that the absolute figures from Table 3.2. The highest per capita waste generation in the WEC survey was found in the OECD countries where, on average, each person generated 2.2 kg/day in 2012. Three regions, East and Central Asia, Latin America, and the Caribbean and the Middle East and North Africa, had a per capita generation of 1.1 kg/day. In East Asia and the Pacific, the amount generated per person was 1.0 kg/day; in Africa, it was 0.7 kg/day, and in South Asia, it was 0.5 kg/day.[3] A global waste production map, in this case based on World Bank data, is shown in Fig. 3.1.

Table 3.2 also contains figures for predicted regional waste production in 2025. Between 2012 and 2025, the total global amount of waste generated is expected to rise to 6,069,705 t/day, an increase of 72%. Much of the additional waste production is expected from countries of the developing world. In 2025, the largest regional daily volume of waste production is predicted to be from East Asia and the Pacific region with 1,865,380 t/day, larger than that produced by the OECD countries, 1742,417 t/day. Waste generation in Latin America and the Caribbean is expected to rise to 728,392 t/day; in South Asia, the waste production is predicted to be 567,545 t/day, Africa the level is 441,840 t/day, and in the Middle East and North Africa, it rises to 369,320 t/day.

The figures reflect the way societies are changing. Waste production in the OECD countries is slowing, and its production is expected to peak in 2050. These countries are also doing more to minimize and to exploit their waste. The production of waste will continue to grow

[3]Figures are from the World Bank.

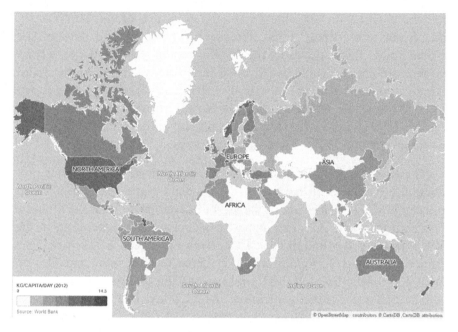

Figure 3.1 Global waste production map. Source: World Economic Forum.

beyond 2050 in other regions and by the end of the century daily waste production could reach 11 million tonnes/day.

While the regional per capita averages quoted above cover a span of per capita waste production of 0.5 to 2.2 kg/day, there are much wider national variations, as shown in Table 3.3. The data in the table, published by the World Economic Forum, indicates that the global average production of waste per person is 1.2 kg/day. However, the highest level of production is found in Trinidad and Tobago, where the average daily waste production per person is 14.4 kg/day or 12 times the global average. As the table shows, island nations make up many of the top waste producers in the table, though none is as high as Trinidad and Tobago. Of the top eight nations in the table, all are small island nations except for Kuwait; all eight produce more than 4.3 kg/day of waste per person. Two other, larger island nations, New Zealand and Ireland, make up the top 10 The top European nation for waste production in the table is Norway with 2.8 kg/day per person, while in the USA, each person produces 2.6 kg/day. At the bottom of the table, meanwhile, are Ghana and Uruguay, each producing only 0.1 kg of waste per individual inhabitant per day. This is 144 times less than the country at the top of the table.

Table 3.3 Waste Production Per Capita Per Day for a Selection of Countries	
Country	Waste Production Per Person Per Day (kg)
World average	1.2
Trinidad and Tobago	14.4
Kuwait	5.7
Antigua and Barbuda	5.5
St Kitts and Nevis	5.5
Guyana	5.3
Barbados	4.8
St Lucia	4.4
Solomon Islands	4.3
New Zealand	3.7
Ireland	3.6
Norway	2.8
Switzerland	2.6
USA	2.6
Ghana	0.1
Uruguay	0.1
Source: World Economic Forum Figures are for 2012.	

Table 3.3 shows how absolute amounts of waste generated vary from country to country. Table 3.4 explores another aspect of waste, how its composition varies from country to country. For low income countries, as already indicated, the highest proportion of waste is organic waste; this accounts for 64% of the total in low income nations. However, for a high income country, the proportion of organic waste is only 28%. For these high income nations, the difference is made up from packaging materials. Paper makes up 31% of the waste in a high income nation, only 5% in a low income nation. Similarly, the proportion of plastic, glass, and metals is higher.

When the World Bank looked at the expected waste composition in 2025 (these figures are not in Table 3.4), it found that the breakdown was virtually identical for the high income nations, although the absolute volume had increased. For the low income economies, the amount of organic waste fell slightly and the proportion of paper and plastic rose. The World Bank study also included lower middle income and upper middle income economies. For both of these categories these the proportion of organic waste was over 50% in 2012 and in 2025, but as with the low income economies the proportion fell during that period, while the amount of packaging materials rose.

Table 3.4 Comparison of Waste Composition from Low Income and High Income Countries		
Material	Proportion of Waste in Low Income Country (%)	Proportion of Waste in High Income Country (%)
Organic	64	28
Other	17	17
Paper	5	31
Plastic	8	11
Glass	3	7
Metal	3	6
Source: World Bank.		

WASTE ENERGY CONTENT

In order for waste to be used in a waste to energy plant, it must have a sufficiently high energy content. According to the WEC, waste can only be considered for combustion if the waste source has an energy content of at least 7 MJ/kg, this being the level at which combustion will be self-sustaining. Of course, there are other ways of deriving energy from waste aside from burning it. Anaerobic digestion can be used to produce a gas from organic waste that might have too low a calorific value for use in a combustion plant. However, the most likely use of MSW in an energy from waste plant is as some form of combustion fuel and so in general the energy content is a key parameter.

Table 3.5 shows the energy content of a range of materials typically found in waste from a variety of different sources, and of a number of fossil fuels. Of the materials found in waste, plastics have the highest energy content, typically around 35 MJ/kg. Textiles with 19 MJ/kg and paper with 16 MJ/kg are also relatively high in calorific value. The typical bag of unsorted waste from a high income level economy, in this case Austria, will have a typical energy content of 8−12 MJ/kg. At this level, the waste would just qualify for incineration by its energy content. In contrast, a typical bag from China would only have a calorific value of 4−5 MJ/kg. Such waste could not be sent to a waste to energy plant without first sorting it and isolating the higher energy content fractions.

Table 3.5 also shows the energy content of other fuels. Wood, which is widely used for heating and cooking in many parts of the world as well as being burnt in biomass fueled power plants, has a

Table 3.5 The Energy Content of Waste and of Other Common Fuels	
Energy Source	Calorific Value (MJ/kg)
Paper	16
Organic material	4
Plastics	35
Textiles	19
Residual waste, unsorted, Austria	8–12
Residual waste, unsorted, China	4–5
Refuse-derived fuel, Germany	13–23
Wood	15
Black coal	29–33
Crude lignite	10
Natural gas	36–50
Diesel	46
Source: World Energy Council.[4]	

typical energy content of 15 MJ/kg when dried although wood as harvested has a calorific value of 10 MJ/kg. This is similar to the energy content of crude lignite. Refuse-derived fuel, a fuel that is produced from sorted waste, has a typical energy content of 13–23 MJ/kg. Black coal contains around 29–33 MJ/kg, and natural gas varies in energy content between 36 and 50 MJ/kg. Diesel contains 46 MJ/kg.

Some indication of regional variations in the energy content of waste are shown in Table 3.5. However there are significant variations, even within high income regions. For example, figures from the US Agency for International Development[5] show that the average energy content of waste in the USA is 10.5 MJ/kg, while in Western Europe it is 7.5 MJ/kg. This is similar to the energy content of waste in Taiwan. For mid-sized Indian cities, in contrast, the energy content is typically 3.3 to 4.6 MJ/kg.

When planning any waste to energy facility, it is important to analyze the waste stream that will form the energy source carefully and over an extended period. For example, the figures for Indian cites cited above suggest that this waste cannot be used in a waste to energy plant

[4]World Energy Council, World Energy Resources: Waste to Energy 2016.
[5]Mining the Urban Waste Stream for Energy: Options, Technological Limitations, and Lessons from the Field, United States Agency for International Development, 1996 (Biomass Energy Systems and Technology Project DHR-5737-A-00-9058-00).

as collected. However, it is important to note that in Indian cities, the waste is often collected by city sweepers and contains large amounts of earth, stones and sand. In Mumbai, the fraction of noncombustible waste of this type can be as high as 30%. Collecting waste in a different way could result is a waste stream that is much more suitable for energy generation. One way of achieving this is to segregate the waste from more affluent parts of a city where the energy content is likely to be higher. Sorting of waste can also be used to improve its energy content.

CHAPTER 4

Waste to Energy Technologies

There are a variety of technologies that can be used to convert waste into energy. The most common method today is to use some form of combustion. Typically, this will be a combustion plant with a simple fixed or moving grate on which the waste is burnt, with the heat generated being captured in a boiler to raise steam for a steam turbine. Plants of this type are often called mass-burn plants. The combustion system may also be based on the more complex fluidized bed design which is well suited to heterogeneous fuels.

There are, additionally, some advanced methods combustion methods for deriving energy from waste. The two most important are pyrolysis and gasification. Pyrolysis involves heating the waste material in the absence of oxygen. Depending on the temperature to which the waste is heated a range of solid, liquid, and gaseous products are formed. These can be used as fuels for different processes, but some of the product must be used to raise the heat needed to drive the pyrolysis process. Gasification, meanwhile involves partial combustion of the waste in a limited supply of oxygen to produce a combustible gas as a product while leaving an ash residue. This residue may also be combustible.

Within the broad outline above, power-from-waste plants vary enormously. Much depends on the waste to be burnt, its energy content, the amount of recyclable material or metal it contains and its moisture content. Waste may be sorted before combustion or it may be burnt as received. Emission control systems will vary too, with toxic metals and dioxins a particular target, but nitrogen oxides, sulfur dioxide, other acidic gases and carbon monoxide emissions must all fall below local limits. Carbon dioxide emissions may need monitoring to comply with greenhouse gas emission regulations.

Once the waste has been burnt, residues remain. Power-from-waste plants will generally reduce the volume of waste to around 10% of its original amount. A way must then be found to dispose of this residual

Energy from Waste. DOI: http://dx.doi.org/10.1016/B978-0-08-101042-6.00004-2

ash. If it is sufficiently benign, it may be used as aggregate for road construction. Otherwise, it will probably be buried in a landfill. Residues from emission control systems will have to be buried in controlled landfill sites too.

Northern Europe has been the traditional home of waste incineration plants for power generation and it continues to house the largest concentration of such plants. Altogether there were around 440 waste to energy plants in the European Union producing 30 TW h of electricity and 55 TW h of heat in 2009[1]. Japan has also made extensive use of waste combustion, though not always for power generation with around 100 plants in operation while the USA has a similar number. In 2011, there were over 1000 waste-to-energy plants in operation in 40 countries around the world[2]. These plants were estimated to have treated 11% of municipal solid waste (MSW) generated globally.

Europe has also developed the most widely-used waste combustion technology based on waste incineration. Two companies, Martin GmbH based in Munich and the Zurich company Von Roll (now Hitachi Zosen Inova), accounted for close to 70% of the market for the dominant technology called mass burn at the end of the 20th century[3]. The rest of the market is divided among a number of smaller companies, most based in either Europe, the USA or Japan. The dominant European technology has also been widely licensed. It was the source of the technology used in most US power-from-waste plants built in the late 1970s and early 1980s. More recently several developing countries of Asia have taken interest in power-from-waste and European technology has been modified for use in China, for example. At the same time the newer technologies based on gasification and pyrolysis are being developed by a variety of companies. These are based on technologies from other industries such as petrochemicals.

For agricultural waste, simpler waste to energy plants based on biomass combustion technology are normally suitable. Agricultural wastes are usually classed as biomass and they will attract renewable

[1] Figures are from the Confederation of European Waste to Energy Plants.
[2] Review of State-of-the-Art Waste-to-Energy Technologies: Stage Two—Case Studies, prepared by WSP Environmental Ltd for the Waste Management Branch of the Department of Environment and Conservation, Australia, 2013.
[3] An overview of the global waste to energy industry, Nickolas J Themelis, Waste Management World, July—August 2003.

subsidies. The technology for burning biomass is very similar to that used for coal-fired power plants but at a smaller scale. Emission control systems are simpler than for plants burning MSW because there are unlikely to be any toxic emissions. Agricultural waste can also be mixed with coal and burnt in a conventional pulverized coal-fired power plant, a process called cofiring. In addition, refuse-derived fuel (RDF) and solid-recovered fuel (SRF), both of which are types of fuel that has been made from sorted waste, can be burnt in conventional power plants as well as being used by other industries such as cement manufacturing.

While combustion technologies dominate, there are other methods of generating energy from waste. One of the most important is anaerobic digestion. This is a process in which bacteria break down organic material in the absence of oxygen, producing a gas that is rich in methane. Small-scale anaerobic plants are used to process animal wastes on large dairy and pig farms across the world. However, the most common type of anaerobic digestion takes place in waste landfill sites where organic material has been buried. These sites generate large volumes of methane which is captured and can be used to generate electric power.

MASS BURN TECHNOLOGIES

The traditional method of converting waste to energy is by burning it directly in a special combustion chamber and grate, a process which is often called mass burning. The flow diagram for a mass burn plant is shown in Fig. 4.1. The dominant European technologies use this system. These involve specially developed moving grates which are often inclined so that waste moves across the grate under the force of gravity. They are also designed so that the waste spends a long time within the combustion zone to ensure that the waste is completely destroyed. Designs for this type of plant have evolved over 20–30 years and are generally conservative, based on traditional technology.

More recently, fluidized bed combustion systems have sometimes been used in place of traditional grates. Such systems are good at burning heterogeneous fuel but require it to be reduced to small particles first.

The actual grate, be it conventional or fluidized bed, forms only a part of a waste treatment plant. A typical solid waste combustion

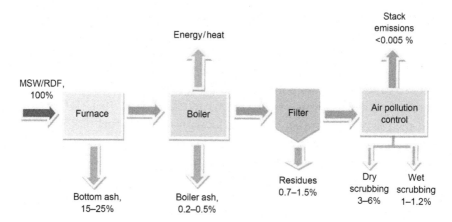

Figure 4.1 Waste to energy mass burn plant flow diagram. Source: Waste Control database of waste management technologies.

facility is integrated into a waste collection infrastructure. Waste is delivered by the collecting trucks to a handling (and possibly a sorting) facility where it must be stored in a controlled environment to prevent pollution. Recyclable materials may be removed at this stage, though metallic material may be recovered from the ash residue after combustion. Grabs and conveyors will then be used to transfer the combustible waste from the store to the combustor.

Plant components, and particularly the grates, must be made of special corrosion resistant materials. The grate must also include a sophisticated combustion control system to ensure steady and reliable combustion while the quality and energy content of the refuse fuel varies. In some more modern systems oxygen is fed into the grate to help control combustion. The temperature at which the combustion takes place must usually be above 1000°C to destroy chemicals such as dioxins but must not exceed 1300°C as this can affect the way ash is formed and its content.

Hot combustion gases from the grate flow vertically into a boiler where the heat is captured to generate steam. The combustion process in the grate and the temperature profiles within the boiler have to be maintained carefully in order to control the destruction of toxic chemicals. Most of the residual material left after combustion is removed from the bottom of the combustion chamber as slag. However there may be further solid particles in the flue gases, some of which can be recycled into the furnace.

Upon exiting the combustion and boiler system, the exhaust gases have to be treated extensively. While the combustion chamber may utilize techniques to minimize nitrogen oxide emissions—though further reduction may prove necessary—a system to capture sulfur will be required. This will probably be designed to capture other acidic gases such as hydrogen chloride too. There may be a further capture system based on active carbon which will absorb a variety of metallic and organic residues in the flue gases. Then some sort of particle filter will be needed to remove solid particles being carried along with the flue gases. By this stage the exhaust gases should be sufficiently clean to release into the atmosphere but continuous monitoring systems are required to make sure emission standards are maintained. The fear of toxic emissions has often been the source of local objections to the construction of waste to energy plants, so to gain acceptance plants have to be scrupulous in their operations.

Dust from the flue gas filters is normally toxic and must be secured in a landfill. Other flue gas treatment residues will probably need to be buried too. The slag from the combustor may, however, be clean enough to exploit for road construction. Modern mass burn plants aim to generate slag that can be utilized in this way.

AGRICULTURAL WASTES

When crops are harvested for food or other uses there is normally a residual waste material. Typical examples are straw left when cereal crops are harvested, rice husks from rice, shells from coconuts, and bagasse, the residue from sugar cane harvesting. Forestry also produces similar residual waste although this is often much more difficult to collect.

Agricultural residues are produced in large volumes across the globe and all these wastes are potential sources of energy. Some, such as sugar cane bagasse, have been used for many years to generate heat and sometimes electricity. Others have been burnt without energy capture or dumped and allowed to decompose. In recent years, however, greater attention has been paid to their potential and more of these residues are now used for energy production.

These materials generally have a high energy content and can be used as fuel for a combustion plant. Such plants are usually relatively small and the efficiency is lower than that of a large steam generating plant. Agricultural wastes can also be mixed with coal as fuel in a conventional coal plant, where higher efficiency is then gained.

ANAEROBIC DIGESTION AND LANDFILL GAS

Anaerobic digestion is a process by which organic material decomposes under the influence of microorganisms in the absence of oxygen. This process occurs naturally in soil and in lakes. The same process can be harnessed in special digesters that are used to treat waste material. In each case, the product of the digestion is a mixture of gases, usually with methane as the main component. This can be used to generate electricity and heat.

When municipal waste is buried in landfill sites, methane is a common product, the gas resulting from natural anaerobic digestion caused by soil bacteria. This gas is normally collected today and used for energy production. Similar processes occur naturally. Marsh gas, for example, is a methane rich gas that results from anaerobic digestion and the same gas can be produced in the bottom of lakes when organic material is carried down rivers and collects in the body of water. Manmade reservoirs can also be a large source of methane if not managed correctly.

Anaerobic digestion can be used to treat waste materials. It is often used in sewage works that treat human waste and a similar process can be used to convert animal waste on farms into a useful energy product. These types of application are normally small scale.

GASIFICATION AND PYROLYSIS

In recent years, a number of companies have developed new waste-to-energy technologies based on both gasification and pyrolysis. These technologies are derived from the power and the petrochemicals industries. Pyrolysis is a partial combustion process carried out at moderate to high temperatures in the absence of oxygen and it can produce a mixture of gaseous, liquid and solid residues. The traditional method of producing charcoal is a form of pyrolysis as was the production of town gas from coal, leaving a residue of coke. Gasification, meanwhile, involves heating solid material at high temperature in a limited amount of air or oxygen to produce a char waste and a combustible gas. In both cases, the gas will normally be burnt to generate heat and steam.

Pyrolysis can produce a range of products from waste, depending upon the temperature and the time the waste spends in the pyrolysis reactor. At lower temperatures with short residence times, more oils and tars are produced. Longer residence times lead to more solid residue (char). When the temperature is low, these solid and liquid products can contain complex and sometimes toxic organic molecules and must generally undergo combustion at a controlled high temperature to generate power in a conventional boiler system.

Many waste to energy pyrolysis plants operate a relatively high temperatures so that they do not produce any liquid or tar residues, only combustible gas and a solid residue. Waste is normally sorted first, removing and recycling as much metal, glass and plastics as possible. The remainder is then shredded and reduced to small particles before exposing it to a high temperature to convert it very quickly into a combustible gas and solid ash. The gas is cleaned and can then be used in a gas engine to generate power. Alternatively, the gas can be used in a conventional boiler.

Waste gasification is similar to pyrolysis, but it generally takes place at a higher temperature to produce a combustible, low calorific value gas that can be burned either in a gas engine or in a conventional boiler system. As with pyrolysis, the products of gasification depend upon the temperature used and lower temperatures can lead to more contaminants in the gas. The synthetic gas produced during gasification is generally a mixture of carbon monoxide and hydrogen but the carbon monoxide can be converted into more hydrogen by using a second reaction in which the gas is mixed with water vapor and passed over a catalyst at high temperature. Provided it is clean enough, this gas can be used either for power generation or as a feedstock for industrial processes.

A new version of the gasification process that is being developed for waste processing is plasma gasification. This involves burning the waste in a plasma arc at very high temperatures. The temperature is so high that all the components of the waste are broken down to atomic constituents and the product is expected to be a relatively pure syngas, a mixture of hydrogen and carbon monoxide.

REFUSE-DERIVED FUELS

The treatments for waste outlined above generally involve converting waste to energy in a dedicated waste to energy plant. There is an alternative. Waste can be sorted and then converted into a form of fuel that is suitable for use in conventional power plants and for industrial use. RDF is a fuel of this type, as is the more carefully controlled, SRF.

RDF is the product of a process to treat MSW to create a fuel that can be burnt easily in a combustion boiler. In order to produce RDF, waste must be shredded and carefully sorted to remove all noncombustible material such as glass, metal, and stone. Shredding and separating is carried out using a series of mechanical processes which are energy intensive. The World Bank has estimated that it requires 80–100 kW h to process 1 t of MSW and a further 110–130 kW h to dry the waste[4].

After the waste has been shredded and separated, the combustible portion is formed into pellets which can be sold a fuel. The original intention of this process was to generate a fuel suitable for mixing with coal in coal-fired power plants. This, however, led to system problems and the modern strategy is to burn the fuel in specially designed power plants. An alternative is to mix the RDF with biomass waste and then burn the mixture in a power plant. Since RDF production must be preceded by careful sorting, this type of procedure is best suited to situations where extensive recycling is planned.

SRF is similar to RDF but with a much more closely defined and controlled specification. The specification is defined by a set of European Standards (CEN/TC 343). The specification of a typical commercial SRF is shown in Table 4.1.

The specification for SRF includes its calorific value, moisture content and the amount of specific pollutants it may contain. Fuel produced to this specification is more predictable in its behavior than RDF and can more easily be utilized by industries. SRF is used in industrial processes such as the production of cement and lime and can be used for heat and power generation too. SRF may be formed into

[4]Mining the Urban Waste Stream for Energy: Options, Technological Limitations, and Lessons from the Field, United States Agency for International Development, 1996 (Biomass Energy Systems and Technology Project DHR-5737-A-00-9058-00).

Table 4.1 Typical Solid Recovered Fuel Specification	
Particle Size	Typically Less Than 35 mm
Calorific value	17 to 22 MJ/kg
Moisture content	Less than 15% by weight
Chlorine content	Less than 0.9% by weight
Sulfur content	Less than 0.5% by weight
Source: Sita.[5]	

pellets, like RDF or it may be delivered in bales or loose, depending upon the requirements of the end user. Both RDF and SRF can often be exported whereas many jurisdictions do not allow the export of unsorted waste.

[5]A guide to solid recovered fuel, Sita, United Kingdom.

CHAPTER 5

Landfill Waste Disposal, Anaerobic Digestion, and Energy Production

The production of a combustible gas from organic waste material is a natural process of decomposition called anaerobic digestion that takes place in soil and water, under suitable conditions. This same process can be exploited to generate a methane rich gas from certain anthropomorphic wastes.

The generation of an energy rich gas from waste in this way is typically carried out at a relatively small scale using animal or human wastes as the feedstock. This is a relatively low technology means of generating energy that is typically used by sewage farms, dairy and pig farms, and by chicken farms. Depending on the size of the plant the energy will generally be used locally. However, there are some larger plants—often based around a regional system of waste collection that is capable of supplying a large volume of feedstock—that deliver their power to the grid.

Although small-scale anaerobic digestion is used for waste conversion in dedicated biogas generating systems, the main source of biogas from anthropomorphic waste is that produced by landfill waste sites. The gas is an important source of global warming. According to estimates from the Global Methane Initiative (GMI), landfill methane is the third largest anthropomorphic source of atmospheric methane, globally, with emissions of close to 800 m tonnes CO_2 equivalent in 2010. These emissions were estimated to be 11% of total global methane emissions from all sources that year.[1] Methane is a much more potent greenhouse gas than carbon dioxide but dissipates more quickly and the conversion to CO_2 equivalent takes account of that. The 2010 annual emissions of 800 m tonnes CO_2 equivalent equates to around 32 m tonnes of methane. Note, though, that estimating the amount of

---

[1]The two leading sources of atmospheric methane production are enteric fermentation from animals such as cattle and fugitive losses from the oil and gas industries.

Energy from Waste. DOI: http://dx.doi.org/10.1016/B978-0-08-101042-6.00005-4

gas released is difficult and annual emission from landfill could be as high as 70 m tonnes/year.

Methane production in landfill sites takes place as a result of bacteria in soil interacting with organic material in the buried waste. In the past, this gas has simply escaped into the atmosphere but more and more sites are capturing the gas and using it to generate either heat or electric power or both. In some parts of the world legislation to combat global warming now requires this.

LANDFILL GAS

Landfill gas is the name given to the energy rich gas generated when organic material is buried in a landfill waste site. The amount of gas produced and its value as an energy source depends on the structure of the landfill. More advanced, and more sanitary, landfill sites will generally bury waste quickly and effectively, creating better conditions for the production of the energy rich landfill gas. Less well managed and maintained landfills are less productive in this respect.

The gases which are released by a landfill waste site have three sources. The most important of these is bacterial decomposition in which the organic waste that has been buried under a layer of soil is broken down by bacteria within the soil. The bacterial decomposition involves a complex series of processes. There are four key stages, each involving a different type of bacteria. These are shown in Fig. 5.1. These stages are called hydrolysis, acidogenesis, acetogenesis, and methanogenesis. During the hydrolysis stage, specialist bacteria break down large organic molecules, such as carbohydrates into smaller constituents such as sugars. This is followed by acidogenesis in which further breakdown occurs in a process similar to the souring of milk. The third stage, acetogenesis, leads to the production of mainly acetic acid by a third set of bacteria. Finally, methanogenesis is carried out by bacteria that convert the products of the preceding stages into a mixture of methane, carbon dioxide and water. Depending on the composition of the waste, there may be other products such as ammonia and hydrogen sulfide, produced in certain stages too.

The second process that can occur in a landfill site is volatization. This is the result of certain wastes, usually organic compounds, changing from a liquid or solid into a vapor. The products of this process

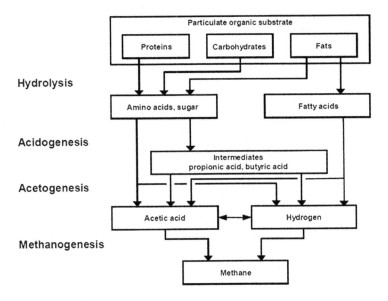

Figure 5.1 Landfill anaerobic digestion process diagram. Source: Waste to Energy Research and Technology Council.

include a range of nonmethane organic compounds. These are usually much more complex organic molecules than methane. They often add to the energy content of the landfill gas. The third source of gases is chemical reactions which can take place between different wastes. These are often chemical wastes such as bleach and can lead to toxic gaseous products. These products are not normally of value as an energy source.

Bacterial decomposition of waste is the most important source of landfill gas. The stages of this process, outlined above, occur over time as conditions within the landfill evolve. Initially, there is oxygen present within the landfill and this allows aerobic bacteria to act on the waste, producing carbon dioxide as they break complex materials down. Once all the oxygen has been used up, these bacteria can no longer survive and anaerobic species take over. These act on the broken-down material, creating acids and alcohols, which make the landfill acidic. These acids, in turn, release other elements from the soil and encourage further types of bacteria to thrive. Eventually, conditions become less acidic and methane producing bacteria can start to operate. This normally occurs within 3 years of the waste being buried. Once the landfill reaches these conditions, it will generally stabilize and it will then continue to produce methane-rich landfill gas for

Table 5.1 Landfill Gas Composition	
Component	Percentage by Volume
Methane	<60
Carbon dioxide	40−60
Nitrogen	2−5
Oxygen	0.1−1
Ammonia	0.1−1
Nonmethane organic compounds	0.01−0.6
Sulfides	0−1
Hydrogen	0−0.2
Carbon monoxide	0−0.2
Source: US Agency for Toxic Substances and Disease Registry.	

around 20 years. This may be extended if the landfill contains a high percentage of organic waste and gas production can continue for 50 years, though not usually at a high level.

Table 5.1 shows a breakdown of the constituents of typical landfill gas as percentages by volume. As the table shows, the two main constituents are methane and carbon dioxide. Methane normally accounts for up to 60% the total, although the proportion can be lower, while carbon dioxide can form from 40% to 60%. Nitrogen will be present at between 2% and 5%. All the other constituents are present in relatively small amounts. These include ammonia which can account for up to 1%, as can sulfides. There are traces (up to 0.2%) of hydrogen and carbon monoxide, and small amounts of oxygen too. The nitrogen and oxygen will normally be from the air.

Table 5.2 contains figures for the atmospheric emissions of landfill methane from ten countries that are part of a project called the GMI which aims to reduce anthropomorphic methane emissions. The ten included in the table are the top emitters of all the GMI nations. All the emissions in the table have been converted into CO_2 equivalent. Top of this list, with 129.7 m tonnes CO_2 equivalent is the USA. China, the second nation in the table had emissions of 47.1 m tonnes CO_2 equivalent or well under half those of the USA, while India's emissions, at 15.9 m tonnes CO_2 equivalent, are less than one eighth those of the USA. Mexico, Russia, Turkey, and Indonesia all had relatively high emissions in 2010, as did the United Kingdom and Canada.

| Table 5.2 Landfill Methane Emissions from Top Ten Countries in Global Methane Initiative ||
Country	Landfill Methane Emissions in 2010 (m tonnes CO_2 Equivalent)
United States	129.7
China	47.1
Mexico	38.4
Russia	37.1
Turkey	33.1
Indonesia	28.3
Canada	20.7
United Kingdom	18.9
Brazil	17.8
India	15.9
Source: Global Methane Initiative.	

The largest landfill methane emissions are generally found in the most developed countries since these are the countries that produce the most waste. However the way in which waste is managed also plays its part and some developed countries are making a much greater effort to control emissions than others. Meanwhile, the waste production is some of the developing nations, such as China and Brazil, is increasing and emissions from these and others are likely to increase in coming years.

ENERGY FROM LANDFILL GAS

Landfill gas has an energy content of around half that of natural gas, making it a valuable energy source for power or heat and power production. In order to be of value, the gas must first be collected from the landfill. This is achieved by inserting a series of wells into the landfill burial site. If gas collection is planned at the start of a landfill, then the pipework can be installed as the waste is dumped. However in many cases a landfill has already been established. In this case the wells must be inserted by boring down into the body of the waste site.

The gas collection system consists of vertical, perforated pipes, often of plastic construction, that are sunk into the landfill. Where it is possible, there may be horizontal pipes added as well to create an interconnected network within the body of the landfill. All the pipes are connected to an above-ground collection system from which the gas is

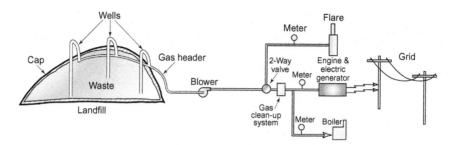

Figure 5.2 Landfill gas recovery and exploitation schematic. Source: The Global Methane Initiative.[2]

extracted for use. The collection system is attached to a pump that places the underground collection wells under a slight negative pressure. This causes the landfill gas to migrate towards the collectors from which it is extracted. The gas that is collected will contain water vapor that must be removed before the gas can be used for energy production. It must also be cleaned to remove any toxic or corrosive gases such as hydrogen sulfide or ammonia that might otherwise damage power generation units. A simplified diagram of a typical landfill gas collection system is shown in Fig. 5.2.

Once the gas is above ground and cleaned, it can then be used as an energy source. There are various options depending upon the location of the site and the local demand. The most common approach is to use the gas to generate electricity. This can be carried out using piston engines, gas turbines, micro turbines, and fuel cells. However, the most common type of power generation uses a piston engine that is adapted to burn natural gas. Typical is a landfill outside Athens, Greece, which has 11 gas engine cogeneration modules, each capable of generating 1.225 MW of power which is fed into the local grid. The modules also generate around 16.5 MW_{th} of thermal energy, most of which is used to generate steam and hot water.[3]

There are other uses for the gas in place of electricity generation. One is to use the gas directly for some industrial process, usually generating heat or steam. It can be used in boilers, dryers, kilns, or to heat greenhouses. Alternatively, the gas can be processed so that it can

[2]Figure is taken from a PDF entitled: Landfill Methane: Reducing Emissions, Advancing Recovery and Use Opportunities.
[3]Utilization of Landfill gas for Energy Production—Operational Experience from a 13.8 MWe Power Plant, G Skodras, P S Amarantos, E Papadopoulou and E Kakaras, EU Organisations for the Promotion of Energy Technologies (OPET) Network.

be supplied directly into natural gas pipelines or be used to provide liquefied natural gas for vehicles.

ANAEROBIC DIGESTION OF ANIMAL FEEDSTOCKS

The natural process of organic fermentation to produce methane can be harnessed to process agricultural wastes and use them to provide an energy source. While, in principle, this can be applied to many sorts of waste, it has traditionally been carried out using animal wastes from high intensity livestock farms. More recently mixed animal and crop wastes have been fermented together, with the combination providing a higher energy output.

The use of fermentation to process farm animal waste serves a dual purpose. It generates a useful energy output and it provides a way of neutralizing the waste material which is often considered hazardous to the environment unless treated. Typically, waste digesters for animal wastes have been used on large dairy and pig farms but they are not generally considered economical for smaller farms to operate even when the value of the waste treatment is taken into account.

An alternative is to establish a large fermentation plant to serve a region where there are a number of small farms located. In countries such as Denmark, plants of this type have been set up as cooperatives that are owned and shared by the farms that use them. Elsewhere commercial anaerobic digestion plants have been established on a commercial basis, charging a fee for the taking of waste. Larger plants of this type might process both animal and crop waste as well as food waste, usually from sites within a 10-km radius.

Anaerobic digestion of animal wastes is a relatively inefficient process. A waste throughput of 2500 m^3/day would only produce around 3–4 MW of power. Using a mixture of animal slurry and green crop silage can improve the efficiency of digestion and so make plant operation more economical. For many plants a typical feedstock will be 70% green silage and 30% animal slurry.[4] Economics can also be improved for larger plants if they can charge a gate fee for the disposal of waste.

[4]Anaerobic Digestion across the United Kingdom and Europe, Suzie Cave, Northern Ireland Assembly Research and Information Service, 2013.

Anaerobic digestion is used widely in some countries in Europe. In Germany for example, there were around 7215 plants operating at the end of 2011 with a total generating capacity of 2904 MW. Many of these plants take advantage of renewable feed-in tariffs to sell power to the German grid. Denmark also makes extensive use of anaerobic digestion, usually in collective plants that generate gas from a mixture or agricultural and food wastes. Other countries such as the United Kingdom have made little use of the technology in the past, although interest is growing. In the United States, there were around 200 anaerobic digestion systems operating in 2015, mostly small-scale, single farm units.

BIOMASS DIGESTERS

As already noted, biomass digestion is generally only cost effective for large farming operations, most usually on dairy or pig farms where the slurry produced by the animals must be treated to prevent it causing an environmental hazard. Digesters of differing sophistication are available depending on the size of the farm. For small farms, the most suitable is usually what is called a lagoon digester, essentially a pond (the lagoon) into which the slurry is placed. The lagoon is covered with an impermeable membrane which is used to contain and collect the emitted gas. The slurry should contain less than 2% solids and the lagoon must usually be maintained above 30°C which limits the application to warmer climates since it is not economical to heat a lagoon digester. The waste must needs to be left in the lagoon digester for from 20 to 150 days in order for digestion to go to completion. The liquid residues can often be used to fertilize fields.

A more sophisticated system is the tank digester. Slurry is loaded into a tank that is fitted with a stirring mechanism to mix the contents evenly (see Fig. 5.3). The tank can be heated to keep the fermentation at the optimum temperature. Tank digesters can handle slurries with 3% to 10% solids. For slurries with higher solids content a plug flow digester is preferable. This has three elements, a mixing tank, a digester tank, and a settling tank. The slurry is first fed into the mixing tank from where it enters the digester tank which contains heating pipes to maintain the ideal temperature. The material moves slowly across the digester tank, with fermentation proceeding to completion in about 20 days as it crosses the tank. After 20 days, it passes into a settling tank

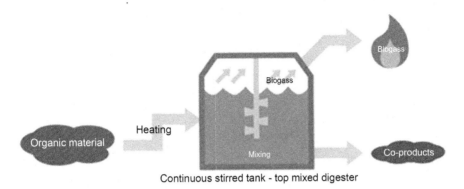

Continuous stirred tank - top mixed digester

Figure 5.3 Anaerobic tank digester. Source: American Biogas Council.

where the remaining solid material is removed and can be used as fertilizer.

The gas from an anaerobic digester has a heating content of 22 MJ/m^3, suitable be burnt in a reciprocating engine to generate electricity and heat. (Natural gas contains around 37 MJ/m^3.) However the capital cost of such systems is high and can only be supported when there is a large quantity of waste to ferment. Similar systems can be used to treat municipal sewage waste and they form an effective means of both rendering it harmless and producing a valuable by-product. Most biomass digester-based power generation plants are relatively small with capacities of 100 kW or less. A new generation of commercial anaerobic digestion plants designed to process a range of material including food waste as well as farm wastes have generating capacities in the 1−10 MW range.

CHAPTER 6

Traditional Waste Combustion Technologies

The traditional method of converting waste-to-energy is by burning it directly in a special combustion chamber and grate. Plants of this type became popular in many countries in Europe during the twentieth century. The dominant European technologies have also formed the basis for designs that have been adopted in other parts of the globe. Often called mass-burn plants, these waste-to-energy plants have specially developed moving grates, often inclined to help the transfer of the waste over the combustion zone, and long-combustion times to ensure that the waste is completely destroyed.

The largest mass-burn plants may burn up to 1200 t/day of municipal solid waste (MSW) on a single grate. For larger capacity more, parallel grates can be added. There are also some large rotating combustor plants, similar to the moving grate in operation. Where a much smaller waste handling capacity is required, a simpler type of combustion system, called a rotary kiln, can be employed. As its name suggests, this system also uses a rotating combustion chamber. Such combustors are capable of burning waste with a high-moisture content, perhaps up to 65%. More recently, fluidized-bed combustion systems have also been used in place of traditional grates. Such systems are good at burning heterogeneous waste but require the combustible material to be shredded into small particles first.

Waste contains a range of materials which are difficult to burn and can produce aggressive combustion products. In consequence, plant components, particularly the grates, must be made of special corrosion-resistant materials. Modern plants must also include a sophisticated combustion-control system to ensure steady and reliable combustion, while the quality and energy content of the refuse fuel varies. In some waste combustion systems, oxygen may be fed into the grate to help control the combustion process. The temperature at which the combustion takes place must usually be above 1000°C to destroy chemicals such as dioxins but must not exceed 1300°C as this can affect the way ash is formed, and its composition.

Energy from Waste. DOI: http://dx.doi.org/10.1016/B978-0-08-101042-6.00006-6

Hot combustion gases from the grate flow into a boiler where the heat is captured to generate steam. The combustion process in the grate and the temperature profiles within the boiler have to be maintained carefully in order to control the destruction of toxic chemicals. Most of the residual materials after combustion are removed from the bottom of the combustion chamber as slag. However, there may be further solid particles carried away in the flue gases.

Upon exiting the combustion and boiler system, the exhaust gases have to be treated extensively. Nitrogen oxides produced during combustion will have to be removed as well as any acid gases such as sulfur dioxide or hydrogen chloride released by the process. Other flue gas cleaning processes are needed to capture and remove a variety of metallic and organic residues in the flue gases. Finally, a form of particle filter will be needed to remove the fine solid particles carried over by the flue gases as well as adsorbent materials that may have been added during earlier cleanup processes. By this stage, the exhaust gases should be sufficiently clean to release into the atmosphere but continuous monitoring is required to make sure that emission standards are maintained. A schematic of a waste-to-energy combustion plant is shown in Fig. 6.1.

Dust from the flue gas filters is normally toxic and must be disposed of in a landfill. Other flue gas treatment residues will probably need to be buried too. The slag from the combustor may, however, be clean enough to exploit for road construction. Modern mass-burn plants aim to generate solid residues that can be utilized in this way.

THE WASTE PROCESSING AND TREATMENT FACILITY

A waste-to-energy plant will often form the hub in a municipal waste-management infrastructure. Central to the hub will be the combustion grate, waste heat boilers, the steam turbine generator, and emissions management facilities, but these only form a part of the operation.

The plant will be supplied with waste collected from the local region using trucks that make daily or weekly collections throughout the area. The waste may be sorted before collection but often the collection will be of mixed waste. This will be delivered to the waste plant where it must be stored in giant bunkers which are enclosed, with closely controlled environments to avoid local problems from smells or other intrusive emissions. These bunkers must be able to hold several days' supply of waste if the plant is to run around the clock.

There may be sorting systems at the waste-to-energy plant that can isolate and remove recyclable material. This will then be transported elsewhere for reuse. The remaining waste will be mixed and perhaps shredded before being burnt. Once combustion of the waste has been carried out, the various residues from the combustion process must then be taken away, some for reuse, others for disposal in the refuge of last resort, a landfill site.

All these operations must be integrated so that waste can be delivered, sorted, burnt, and the residues disposed of continuously. Modern life requires this process to run smoothly week after week and month after month. Any interruption to the waste-collection procedure can lead to a major social problem.

MOVING GRATE WASTE-TO-ENERGY PLANTS

The mass-burn combustion plant is the most common type of waste-to-energy plant. These plants are extremely flexible and can burn waste with varying composition and energy content. In the second decade of the 21st century, over 1000 plants of this type were operating worldwide. The term "mass-burn" was originally used for plants that simply burnt municipal waste to reduce its volume. This practice is not considered environmentally attractive today, and in consequence, mass-burn plants are normally waste-to-energy plants. Although the overall disposal philosophy has changed, many of the earlier waste-combustion technologies are still utilized although technical advances have rendered them more effective.

Mass-burn plants burn mixed MSW that will come from a variety of sources and contain different materials with a range of calorific values and differing moisture contents. The waste may already have been sorted before being delivered to the plant so that it does not contain incombustible fractions such as metals and glass. If not, some sorting may take place on site. Alternatively, metals may sometimes be recovered from the ash once combustion is complete.

The MSW is usually stored in a bunker that can hold material sufficient for several days' supply so that the plant can operate continuously. Waste is normally delivered only during conventional working hours on the five working days each week. In the bunkers, cranes may be used to move and to mix the waste in order to try and create a more homogeneous fuel for the waste-to-energy system. Waste may, additionally, be shredded in the bunker to make combustion easier. From the bunker, prepared waste will be loaded into a feed hopper and chute that delivers the waste to the combustion grate. Upon exiting the feed chute, the waste is normally loaded onto the combustion grate using hydraulic rams. In a modern plant, the operation of these will be controlled using an automated system that meters the waste delivery to ensure uniform loading of waste onto the combustion grate.

The dominant mass-burn grate technology is called a moving grate combustion system. As the name implies, the waste is burnt on a grate that moves in such a way that the burning material is transported across the grate. There are a number of designs for systems of this sort but the most common are reciprocating grates and roller grates. Both are normally inclined to assist motion of the waste material across the grate during the combustion process. There are horizontal designs too.

A reciprocating grate consists of sets of bars perpendicular to the direction of motion of the waste across the grate. These bars are driven hydraulically in a cyclical motion that slowly moves the waste material from one bar to the next as it burns. The alternative roller grate replaces the bars with rollers across which the waste moves as the rollers turn. Whichever type, the grate will usually be sloping so that gravity also helps move the material across the grate. The high temperature in the combustion zone means that the grates are exposed to high temperatures too and the main grate elements are usually cooled, either using water or air depending on the temperature zone. Moving grates may have both air-cooled and water-cooled elements.

A common configuration for a mass-burn plant is a called a waterfall combustor. In this configuration, there will be two or more inclined grate modules, one following the other. Often there will be three modules; the first, called the drying grate, reduces the moisture content of the waste using heat from later stages. The second, burning grate is where most of the combustion takes place. The third, finishing grate, ensures that all combustible material is burnt so that no potential pollutants remain. A schematic of this type of combustion system

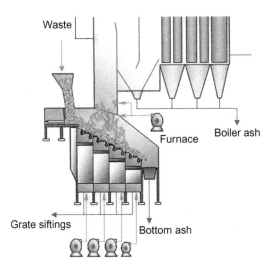

Figure 6.1 A waterfall combustion system. Source: Igniss Energy.

is shown in Fig. 6.1. The ash remaining at the end of this final grate—as well as ash which drops through the grates as the waste moves across the combustor—is discharged into a residual ash management system. This is commonly a quenching pit that cools the ash with water before removal of the moist ash by conveyor. Dry ash systems are also used, but less commonly. Depending on the level of toxicity of the residual ash, it may be used as a building or road material, or it may have to be buried in a landfill.

Combustion of waste in a moving grate combustor takes place in air. Primary air for the combustion is fed from beneath the grate (underfire air). In modern plants, this air can be controlled depending on the waste quality and type. If there are two or three grate modules, each will have its own, independently controlled underfire air supply. The temperature will be controlled within the grates to ensure that all potential toxic materials are destroyed. This will often require a specific transit time through the high temperature zone of the grate.

During combustion, there will be the release of volatile materials. The combustion of these takes place in the zone above the grate, where more air (secondary and tertiary air) is injected. Regulations typically demand that a temperature of 850°C or higher is maintained for two seconds beyond the last air injection point to ensure all potentially toxic material is destroyed.

The largest grates of this type can handle up to 1200 t of waste each day, or over 400,000 t/y. For larger throughputs, several independent lines of grates are used. One of the largest plants of this type in the world, the Afval Energie Bedfijf plant in Amsterdam, the Netherlands, has six incineration lines and can handle 1370,000 t of waste each year.

Energy recovery in this type of plant starts with the walls enclosing the combustion grate. These are built from tubes carrying pressurized water that circulates through the pipes, capturing heat which is carried away. The lower regions of this waterwall combustion chamber are often lined with a refractory ceramic material that is corrosion resistant since the waste may contain a range of corrosive materials. The flue gases that leave the combustion chamber will then pass through further heat capturing modules including superheaters and economizers that extract as much heat as possible from the waste combustion. Following heat capture, the flues gases are then treated to remove any potential pollutants.

ROTARY COMBUSTORS

Less common than a moving grate is the rotary mass burn waterfall combustor. Like the moving grate design, this type of device can handle a relatively large throughput. Waste is fed into the combustor using a hopper, chute and ram feeder similar to that found in a moving grate plant. However, in this case, the waste tumbles and falls slowly along the inclined cylindrical combustor as it slowly rotates. Combustion air is fed into the rotating chamber, with some injected under the waste, and secondary air is introduced above the waste bed. More combustion air is added when the combustion gases exit the rotating combustor. This type of system used less air than the more conventional moving grate, leading to more efficient energy capture.

Heat capture in the rotary waterfall combustor begins within the combustion chamber itself where pipes carrying water pass through the chamber. The rotating system is contained within a combustion chamber with waterwalls, similar to a moving grate design, and these capture more heat energy, while further heat capture elements are placed in the flue gas path.

A smaller rotating combustion system is the rotating kiln combustor. These devices are simpler than moving grate or rotary waterfall

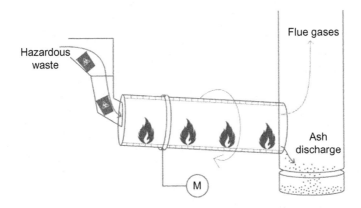

Figure 6.2 Small rotating kiln waste combustion system schematic. Source: Waterleau.

combustors and have no energy capture within the kiln. This makes them cheaper to construct. Some may operate in batch mode rather than being continuously fed with waste. A schematic of this type of combustor is shown in Fig. 6.2.

Rotating kiln waste combustion systems are often used for more difficult or hazardous wastes. They can operate at 800 to 1400°C and some can handle liquids and sludges as well as solids. Energy recovery is possible using a waste heat boiler that captures heat from the flue gases exiting the kiln combustion system. Rotating kiln waste-to-energy plants can have capacities of up to 200 t/day of waste but the smallest, for specialist applications, may have capacities as small as 3 t/day. Large rotating kiln plants may be suitable for small communities while the smallest can be used for treating medical or special industrial waste on-site.

FLUIDIZED BED COMBUSTORS

A fluidized bed combustion system is a special type of combustor that was originally developed as a reactor for the chemical process industry. The combustor exploits the rapid mixing properties of a fluid, converting a mass of solid particles into a "fluid" by pumping air through the bed of particles from below. This agitation of the bed causes the particles to become "fluidized" so that they mix rapidly with one another and with the air injected into the bed, conditions that accelerate reactions between particles as well as combustion within the bed. This type of combustor can burn a wider range of combustible material than can most conventional combustion systems and it is used for a variety of applications including biomass combustion and for burning poor quality fossil fuels such as peat.

The actual fluidized bed comprises a bed of inert, non-combustible material—usually sand—which is fluidized by pumping air from below. Combustible material, in this case shredded waste, is added to the fluidized bed through a feed system. The waste may be supplemented with additional fuel such as natural gas during start-up, or to aid the complete combustion of poor quality waste. Once combustion has started it will normally continue unaided so long as sufficient fuel is added to the bed.

There are two primary types of fluidized bed that have been used for waste combustion, a bubbling bed combustor and a circulating fluidized bed combustor. The bubbling bed is the classic type of fluidized bed in which the pressure and volume of air pumped through the bed causes the solid particles within the bed to behave like a boiling liquid. This bubbling or boiling leads to extremely good mixing within the bed itself, promoting even, homogeneous combustion. In the bubbling bed the solid particles remain at the base of the combustion chamber and there is a clear region above into which secondary air can be introduced to ensure complete combustion of any volatile material that does not burn within the bed. The even combustion conditions mean that the bed temperature can be maintained at a relatively low 815°C and less air is required than for a moving grate combustor. This improves efficiency of energy capture (Fig. 6.3).

The second type of fluidized bed is the circulating fluidized bed. This uses a higher air flow through the bed so that a significant part of the material in the bed becomes entrained and flows into the area above the main bed. This solid material is eventually captured in a cyclone filter through which the flue gases pass as they exit the combustion chamber. From the cyclone, the captured particles are returned or "circulated" back into the bed. The circulating bed creates greater turbulence than the boiling bed, so in principle a smaller combustion chamber can be used to achieve a similar throughput.

Both types of fluidized bed plant can incorporate energy capture with pipes carrying water passing through the bed. The combustion

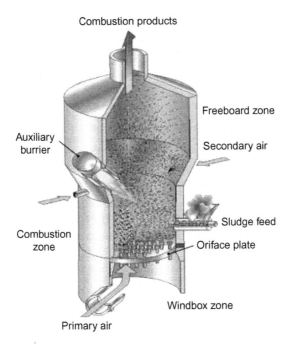

Combustion products

Freeboard zone

Auxiliary
burrier

Secondary air

Sludge feed

Combustion
zone

Oriface plate

Windbox zone

Primary air

Figure 6.3 A bubbling fluidized bed combustion system. Source: Hitachi Zosen Inova.

chamber itself will normally be constructed from waterwalls, as in a moving grate system, and further heat capturing elements will be placed into the flue gas path to maximize energy capture. The waste combustion capacity of a fluidized bed plant for waste combustion varies from as small as 25 t/day to a maximum of around 500 t/day. The technology has been widely used for waste-to-energy plants in Japan where there are more than 80 such plants in operation. There are a limited number elsewhere, with around 10 in Europe.

ELECTRICITY GENERATION

Electricity generation in large mass burn waste-to-energy plants is carried out by raising steam in the combustion chamber and using this steam to drive a steam generator. The combustion temperature in the combustion chamber of a waste-to-energy plant is relatively low compared to typical coal or gas-fired power plants and the steam systems usually rely on traditional designs although some more advanced elements have been introduced into recent plants. In these traditional steam raising boilers, water under pressure passes through tubes in the walls of the combustion chamber (and in some designs within the grate itself) where it is heated to boiling point. The boiling water passes up the waterwall tubes and into a drum above the combustion chamber where water and steam are separated, with water returned through downpipes to the boiler. Steam from the drum is passed into a heat exchanger in the flue gas path called a superheater which increases the temperature of the steam further (superheats it) before the hot fluid enters the steam turbine. Energy is extracted from the steam by blades inside the steam turbine, or turbines, as there may be more than one in a large plant. The steam exiting the steam turbine is then condensed back to water and returned to the boiler, passing as it does through a low temperature heat exchanger called an economizer which captures the last of the heat in the flue gases before they enter the plant emission control systems.

The efficiency of a steam turbine generator depends on the temperature and pressure of the steam. The temperature is particularly crucial for a heat engine because the efficiency depends on the temperature drop across the steam turbine. For a condensing turbine the outlet temperature will fixed at ambient temperature so the only way to increase efficiency is to increase the inlet steam temperature.

For a waste-to-energy plant, the corrosive nature of the flue gases means that both temperatures and pressures must be limited for otherwise excessive corrosion will take place in the energy capture elements such as superheaters in the flue gas path. This inevitably limits the steam temperature and pressure that can be achieved. In consequence, most waste-to-energy plants are relatively inefficient by modern standards.

The actual steam conditions that can be used will depend on the type of waste being burned. For example, in Asia steam conditions have traditionally been limited to 40 bar and 400°C because of the high plastics content and high water content of much of the waste, both of which encourage corrosion. These steam conditions limit efficiency to around 22% to 25%. More recently, in Europe, there has been an attempt to gain higher efficiency by using some of the advanced technologies used in fossil fuel combustion plants. These include more advanced combustion control, and the introduction of steam reheat in multiple turbine plants so that steam exiting a first, high pressure steam turbine is heated once again before entering the next, medium or low pressure turbine. This has allowed efficiencies of up to 30% to be achieved.

Generating capacities of waste-to-energy plants tend to be small compared to conventional power plants. For example, the two newest incineration lines at the Afval Energie Bedrijf plant in Amsterdam[1,2] generate a combined 66 MW at an efficiency of 30%. Most plants have generating capacities smaller than this.

AGRICULTURAL WASTE COMBUSTION

The combustion of agricultural wastes for energy production is common but agricultural crop residues are very different from MSW in terms of their combustibility. They are more consistent in their composition than municipal waste and they generally have a much higher energy content. This makes them much easier to convert into energy. In addition they do not usually contain any components that could produce toxic emissions, making pollution control relatively simple. Although agricultural residues are, strictly, wastes, these materials are generally considered to be biomass and therefore renewable.

Typical wastes from crops include straw from cereal crops and maize, bagasse waste from sugar cane and a variety of wastes from tropical crops such as coconuts or palm oil. All these can be burnt in combustion plant to generate heat and electricity but the size and type of plant will depend on the supply available. Large plants will require a supply of fuel that is available all the year round. This may be provided from a single crop that is continuously harvested. Sugar production from sugar cane often relies on process heat from combustion of bagasse with additional energy being used for electricity production. Alternatively, fuels from different sources may be mixed to provide a reliable energy source. Some plantation energy plants burn sugar

---

[1]Review of State-of-the-Art Waste-to-Energy Technologies: Stage Two—Case Studies, prepared by WSP Environmental Ltd for the Waste Management Branch of the Department of Environment and Conservation, Australia, 2013.
[2]Afval Energie Bedrijf plant in Amsterdam. Two new incineration lines were added in 2007, making a total of six. The two new lines together produce 66 MW and the total plant handles 1,370,000 t of MSW each year. Details are published in Review of State-of-the-Art Waste-to-Energy Technologies: Stage Two—Case Studies, prepared by WSP Environmental Ltd for the Waste Management Branch of the Department of Environment and Conservation, Australia, 2013.

cane bagasse and wood. For other crops and in some regions, it makes more sense to collect the agricultural waste from a large area and store it to provide fuel for a power plant sized so that it can operate throughout the year, with generating capacity based on the regional supply available. For example a cereal-crop straw-burning plant may collect waste from regional farms and store it to provide a year-round supply. Another option is to burn different fuels at different times of the year. Smaller plant may only operate when a source of fuel is available but this mode of operation is not usually economical.

A typical agricultural waste combustion power plant will be a direct fired plant in which the combustible material is burned and the heat released is exploited to raise steam which can be used directly in an industrial process, or to produce electricity. A schematic of this type of plant is shown in Fig. 6.4. Most modern plants use a type of combustion system called a stoker grate which allows fuel to be added and ash to be removed from the grate continuously. Systems of this type are often used for wood waste or mixed biomass waste. Fluidized bed systems can be used to burn agricultural waste too. In addition, it is possible to adopt a system similar to that used in a modern pulverized coal-fired power plant in which the waste is reduced to a fine powder and then injected into boiler through a burner, mixed with air, where it burns in a fireball in the center of a combustion chamber. Particle size and moisture content must be carefully controlled in this type of pulverized fuel plant and not all agricultural wastes can be burnt in this way. There are also very simple agricultural waste combustion systems in which the waste is burned in batches, with the combustor shut down at intervals to remove ash.

In all except the simplest of these combustion systems, the steam raising system will use a combustion chamber built from waterwalls and additional heat capturing elements including superheaters and economizers in the flue gas path. Most agricultural waste-to-energy plants of this type are relatively small and operate at lower temperatures than typical fossil fuel fired power plants. This limits their efficiencies to under 30% and many operate at around 25% efficiency.

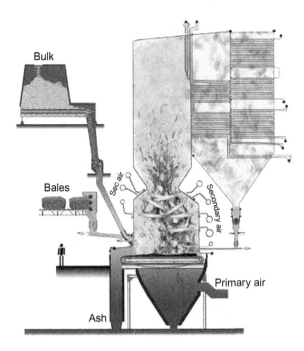

Figure 6.4 Schematic of plant for combustion of biomass fuels such as straw waste. Source: Aalborg Energie Technik.

EMISSION CONTROL SYSTEMS

The complexity of MSW, which usually contains a wide range of different materials, means that its combustion has the potential to generate a range of toxic compounds which could be released into the atmosphere. As a result, waste-to-energy plants are normally required to meet extremely stringent emissions regulations which impose limits on the amounts of different materials that can be released. While these regulations and limits vary from one jurisdiction to another, most cover the same range of potential pollutants and so require similar control systems to be put in place.

The main pollutants that have to be controlled in a waste-to-energy plant are particulates (dust), oxides of nitrogen, acidic gases including sulfur dioxide and hydrogen chloride, heavy metals and some particularly toxic organic compounds including dioxins and furans. There will also be limits on other volatile organic compounds and on the release of carbon monoxide. Each of these is likely to require a separate control system.

A technique that can help control emissions of some of these pollutants is a process called flue gas recirculation. Recirculation involves taking some of the flue gases exiting the combustion chamber and feeding them back into the combustion zone, often as secondary air. This will have the effect of reducing the total amount of nitrogen passing though the combustion chamber, reducing the quantity of nitrogen oxides produced. It also enables some organic compounds in the flue gases to be destroyed during a second pass. Recirculation has the additional advantage of reducing the amount of excess air[3] that is fed into the combustion chamber. The lower the excess air, the more efficient the energy conversion system.

Nitrogen oxides removal is often carried out in two stages. The first is to limit the production of nitrogen oxides (NO_x) in the combustion chamber itself. This is achieved by controlling the combustion zone temperature since NO_x production is greater, the higher the temperature during combustion. Minimization can be assisted by restricting the amount of air in the hottest part of the furnace, encouraging reducing conditions which also help minimize NO_x production. While these techniques can help reduce the production of NO_x they will not normally be sufficient to meet the required regulations alone so further measures are necessary. Two are in common use, selective non-catalytic reduction (SNCR) and selective catalytic reduction (SCR). Both involve adding a reagent to the flue gases, usually ammonia or a urea derivative, that will react with nitrogen oxides, reducing them to nitrogen. With SNCR the reagent is injected into the flue gas stream soon after it leaves the boiler where the temperature is around 850–950°C. Under these conditions, the reagent will react spontaneously with the pollutants in the flue gases. SNCR can achieve a reduction of around 60% in the level of NO_x in the flue gases but can lead to some of the reagent being carried over in the flue gas (a process called ammonia slip). Excess ammonia or urea must be removed at a later stage or it, too, will breach regulations. The alternative is SCR. With this technique the ammonia is injected into the flue gas steam at a much lower temperature, typically between 200 and 300°C. The mixture then passes over a catalytic chamber where special catalysts promote the reaction between the reagent and NO_x. This type of system can remove up to 90% of the NO_x.

---

[3]Excess air is the amount of air about the precise stoichiometric amount needed to convert all the carbon into carbon dioxide and all the hydrogen in the waste into water vapor.

Acid gases are usually treated using a wet or semidry scrubbing system. A wet scrubber comprises a tall tower into the bottom of which the flue gases are injected. Inside the tower there is a system of sprays that fill the space with droplets containing a reagent that will react with any acid gases present, removing them from the gas stream. The most common reagent for a wet scrubbing system is hydrated lime and this technology is the standard way or removing sulfur dioxide from the flue gases in coal-fired power plants. The same technology will also capture other acidic gases such as hydrogen chloride and hydrogen fluoride which might be generated from some types of waste. The reagent, once spent, is removed and depending on its composition may be used as a building material. Semidry and dry scrubbers are similar to wet scrubbers, with a reagent sprayed into the path of the flue gases. However in this case the dry particles become entrained with the flue gases and must be collected downstream in a particle filter.

Dioxins and heavy metals are often carried away in particles in the flue gas stream from a waste-to-energy combustion grate. Metals such as mercury, cadmium and lead can be released into the atmosphere if not captured. The most common method of capturing these, as well as dioxins and furans that may have formed, is to inject activated carbon into the flue gases. Activated carbon consists of very fine particles of carbon which present a large surface area onto which the pollutant materials are preferentially absorbed. The activated carbon is usually injected after a wet scrubber but may be injected into a dry or semidry scrubbing system. The particles are removed later in a particle filter, together with their load of heavy metal and organic compounds.

Particulates are small particles of inorganic material formed or released during the combustion process, particles that are small enough to become entrained and carried away with the flue gases. These particles are classified by their diameters and those smaller than 10 μm are considered the most dangerous because they can be inhaled and enter the lungs. The smaller they become, the more damaging they are considered for human health. While the particles themselves may be inert[4], organic materials and heavy metals may become absorbed onto their surfaces and so carried along with them. Some particles may be removed at the exit of the furnace using a cyclone filter but in most

---

[4]Even though inert, the particles can be extremely damaging if they enter the blood stream.

plants these particles will be removed in a facility called a baghouse filter. This is a series of fabric bags into which the flue gases flow. Particles become trapped within the fabric while the cleaned flue gases pass out the other side. The fabric bag filters can also capture particles from a dry scrubbing system and activated charcoal used for heavy metal and dioxin capture. Fabric filters of this type are considered the most efficient means of trapping fine particles and offer the highest level of capture under the conditions of a waste-to-energy plant. There is an alternative, a device called an electrostatic precipitator (ESP). This comprises a series of wires to which a high voltage is applied and a parallel series of oppositely charged plates. As the flue gas passes through the wires the particles within it become charged and are then attracted to the plates, where they stick. The plates are periodically cleaned by "rapping" them, when the collected dust falls to the bottom of the ESP. This type of dust removal system is effective for larger particles but less so for the smaller particles. As a consequence it is less commonly found in waste-to-energy plants.

CHAPTER 7

Advanced Waste-to-Energy Technologies: Gasification, Pyrolysis, and Plasma Gasification

Although mass-burn technologies continue to dominate in waste-to-energy plants, there are some alternative technologies that have begun to attract attention in the 21st century. These advanced technologies can potentially offer greater efficiency of energy capture, better emission control, and the possibility of generating fuels or synthetic precursors from waste in place of energy. The two main alternatives available are waste gasification and waste pyrolysis. A third, plasma gasification is also being developed for waste-to-energy plants but in terms of deployment it is less advanced than the other two.

The two main technologies differ from mass-burn waste-to-energy in the amount of air or oxygen that is admitted during the process. A mass-burn plant carries out combustion in an excess of air so that all the combustible material is fully oxidized and all the energy is extracted as heat. Gasification, in contrast, is carried out in the presence of a limited amount of oxygen or air. The oxygen allows some combustion to take place but it does not proceed to completion. Instead the waste is converted, by partial oxidation, into a combustible gas that can be used either for electricity generation or as a precursor for the synthesis of chemicals. In waste-to-energy plants, the gas is usually used for power generation.

Pyrolysis of municipal solid waste (MSW) takes place at high temperature in the absence of air. It entails a thermochemical breakdown of the constituents of the waste. The product in this case may be a liquid or a gaseous fuel and a solid char that may also be combustible. The actual products depend on the temperature at which the pyrolysis is carried out, but most waste-to-energy pyrolysis plants use high temperatures in order to destroy any potential toxic materials. Plasma gasification uses an extremely high temperature plasma arc to carry out the gasification of waste which is also performed under depleted oxygen conditions. However, the product is essentially the same as that

Energy from Waste. DOI: http://dx.doi.org/10.1016/B978-0-08-101042-6.00007-8

produced by more conventional gasification. The extreme temperature leads to the complete destruction of the waste feedstock, and there is limited or no production of gases such a sulfur dioxide or nitrogen oxides.

All these technologies have been developed for other industries prior to use in waste-to-energy plants. Pyrolysis is used in the petrochemicals industry, while gasification of coal is considered a promising means of generating electricity with carbon dioxide capture. Plasma gasification has been used in the past to destroy hazardous chemicals and waste. Application of each to waste management has been stimulated by the desire to achieve better energy efficiency during waste disposal.

One of the advantages of these technologies is that they use much higher temperatures than are generally achieved with incineration of MSW. In a waste-to-energy plant that utilises the released energy to generate electricity by generating steam, this can allow the steam cycle to operate at a higher steam temperature leading to greater efficiency. Combustion of the already hot gases produced during a gasification process will also help to generate high temperatures.

GASIFICATION OF WASTE

Gasification of waste is considered attractive because it can convert much of the energy contained in the waste material into a gaseous fuel that has a range of uses. According to one estimate, 80% of the chemical energy within the waste is extracted as a fuel[1] making it potentially much more efficient than incineration. The gas which is produced can be used to manufacture chemicals, but in a waste-to-energy process, it will normally be burnt in a power plant to generate electricity.

Gasification is a process known as partial oxidation and it has been widely used in industry to manufacture hydrogen from fossil fuels. It can be carried out in either air or oxygen, with a partial combustion of the waste material providing the energy needed to drive the process.

[1]Review of State-of-the-Art Waste-to-Energy Technologies: Stage Two—Case Studies, prepared by WSP Environmental Ltd for the Waste Management Branch of the Department of Environment and Conservation, Australia, 2013.

When combustion takes place in excess air—as in an incineration plant—the combustion process, the chemical oxidation of the waste, goes to completion, and all the hydrogen and carbon present in the waste are converted into water vapor and carbon dioxide. During partial oxidation, the amount of oxygen is restricted. Under these conditions, the combustion reactions cannot go to completion and the main products are carbon monoxide and hydrogen.

The reactions that take place during gasification are complex, but there are three main processes which take place. These three processes are the partial combustion of carbon:

$$2C + O_2 = 2CO$$

the water–gas reaction:

$$C + H_2O = CO + H_2$$

and the Boudouard reaction:

$$C + CO_2 = 2CO$$

All these reactions involve carbon contained within the waste material. The reaction is carried out with around 20% of the oxygen that would be needed for the carbon to be completely oxidized. Some complete oxidation does occur and carbon dioxide is formed but the amount is limited.

The partial combustion of carbon and the Boudouard reaction are both exothermic and the energy they release provides the driving force for the water–gas reaction. Taken together, the result of all these reactions is to produce a gas that is primarily a mixture of carbon monoxide and hydrogen with some carbon dioxide. If the source of oxygen is air, then there will also be nitrogen in the mixture. This gas is called synthesis gas or syngas and has been used as a feedstock for a variety of industrial processes. However, since the main components are combustible, so it can also be burned to release energy.

The gasification process can be taken a stage further by reacting the syngas over a catalyst with more steam. This process, called the water-shift reaction, converts the carbon monoxide in the syngas into a mixture of hydrogen and carbon dioxide via the reaction:

$$CO + H_2O = CO_2 + H_2$$

This process is carried out commercially to produce hydrogen, but it is unlikely to be cost effective in a waste-to-energy plant.

There are a range of gasifier reactors that can be used to gasify waste. The main types are fixed-bed gasifiers (updraft and downdraft), moving-bed gasifiers, fluidized-bed gasifiers, cyclonic reactors, and rotary kiln gasifiers. The two most commonly employed are fixed-bed and fluidized-bed gasifiers, but other types may be more suited to specific wastes. Three of the common types are shown in Fig. 7.1.

A fixed-bed gasifier uses a grate within a sealed combustion chamber to support the feed material. As waste is burnt, the residual ash passes through the grate and can be collected from below. There are a variety of fixed-bed gasifiers including a downdraft gasifier, where both the product gas and the solid waste move down through the reactor, an updraft gasifier in which the product gas moves upwards, while the solid waste moves down, and also cross-draft variants. These are all relatively simple devices and it can be difficult to control the temperature within the gasifier, so optimum gasification performance is not always achieved. This makes them less useful for large-scale gasification, and for plants with a generating capacity of over 1 MW.

The main alternative is the fluidized-bed gasifier. This is similar in operation to the fluidized-bed waste-incineration plants described in Chapter 6 in which the waste, first shredded into small particles, is fed

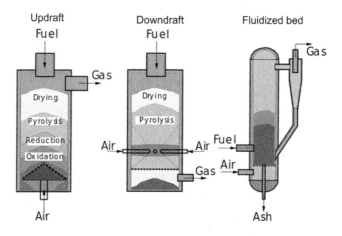

Figure 7.1 Diagrams of three common gasifiers for waste. Source: Wikipedia Commons.

into a bed of sand that acts as the fluidizing medium in which the gasification reaction takes place. The nature of the fluidized bed means that a uniform temperature can be maintained within in the bed and in the region above the bed to ensure that the gasification reactions are completed and that all potential toxic compounds are destroyed. The conditions in the gasification reactor are severely reducing and this helps to prevent the formation of either sulfur dioxide or nitrogen oxides.

Gasification of waste can be carried out either in air or in oxygen. Air gasification is carried out at a temperature of between 900 and 1100°C. Under these conditions, the syngas produced is diluted with up to 60% nitrogen and has a relatively low-energy content of $4-6$ MJ/m^3 (natural gas has an energy content of $37-43$ MJ/m^3). Gasification in oxygen takes place at a higher temperature, $1000-1400$°C and the product has a much higher energy content of $10-18$ MJ/m^3. Although the product is more useful, gasification in oxygen requires an oxygen plant and this adds a significant cost to the waste-to-energy process. On the other hand, high-temperature gasification has the additional advantage that it melts the ash residue, producing an inert slag.

Gasification of waste is popular in Japan because it can be used to reduce the waste volume significantly. High-temperature gasification using oxygen is commonly deployed because this allows any inorganic material within the waste to be melted and converted into an inert slag which can be used as a construction material. The process exploited in Japan is usually called slagging gasification. The oxygen for the slagging gasifier is produced in a special plant, usually by a process called pressure-swing adsorption.

The gas produced during the gasification process is likely to contain a range of impurities although the exact composition will vary with the gasification temperature. Higher temperature gasification leads to fewer toxic contaminants, and in a slagging gasifier, heavy metals can be sealed inside the vitreous slag. Unlike in a mass-burn plant, the gases from the gasifier are not released directly into the atmosphere. However they will still need to be cleaned before they can be used for any other purpose.

In principle, the syngas that is produced in a waste-to-energy gasification plant, once cleaned, can be used as a feedstock for an industrial process. In practice, it is more usual for it to be burnt to produce electricity. Some waste-to-energy plants will burn the gas in a boiler to raise steam that is used to drive a steam turbine. However, it may be more efficient to use the gas in a gas engine or a gas turbine. In both these cases, the gas must be cleaned scrupulously first in order to remove any corrosive gases that might damage to the engines. Whatever the type of plant, the flue gases that emerge from the boiler or from the engines will need to be cleaned so that it meets local regulations regarding emissions of nitrogen oxides, carbon monoxides, particulates, and any volatile organic compounds.

PYROLYSIS OF WASTE

Pyrolysis is the high-temperature decomposition of a material in the absence or air or oxygen. A common traditional example of pyrolysis is the production of charcoal from wood, and a more modern industrial process is the coking of coal to produce a combustible gas from the volatile material it contains and a residue, the coke. The products of pyrolysis depend on the feedstock but also on the temperature at which the decomposition is carried out and the residence time in the reactor. With lower temperatures and shorter residence times, more oils and tars are produced. When both are raised, there is a larger percentage of gaseous product. The low temperature pyrolysis produces more solid residue, char, too, and the liquids are more complex and potentially contain more toxic chemicals that must be destroyed. In most cases, therefore, the pyrolysis of waste takes place at a higher temperature.

If pyrolysis is carried out at above 800°C, the primary product is syngas, similar to that produced from a gasification plant. Since the process is carried out in the absence of air, the product contains a range of hydrocarbon materials as well as both carbon monoxide and hydrogen. The former includes methane and higher hydrocarbons. Smaller quantities of pyrolysis oil may be produced too and a solid char that will contain any inorganic material from the waste as well as some carbon containing materials. A schematic of a pyrolysis process for organic waste is shown in Fig. 7.2.

Reactors that are used for pyrolysis include moving grates reactors, fluidized-bed reactors, rotary kilns, and rotary grates. For high-temperature gasification of waste, the fluidized-bed reactor is popular because it

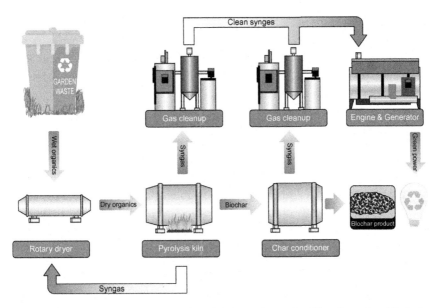

Figure 7.2 Schematic of an organic waste pyrolysis plant. Source: Energy from Waste Research and Technology Council, UK.

can handle the varying composition of the waste easily. Whatever the reactor, however, management of heat transfer is important because the pyrolysis process is highly endothermic and requires significant energy input to proceed.

There are a wide range of MSW pyrolysis plant configurations. One Japanese plant uses two, connected, fluidized-bed reactors. The first carries out the pyrolysis of waste, while the second generates heat from the char remaining after pyrolysis. The two are linked so that the bed material can pass from the first into the second and then be cycled back to the first. Heat for the pyrolysis reaction is provided by hot steam that is pumped through the bed of the pyrolysis reactor. Syngas produced in this reactor is used to generate the steam. Other plants use more conventional arrangements with moving grates or with batch reactors that must be unloaded after each batch of waste has been pyrolysed.

Pyrolysis must normally be carried out using carefully sorted waste. This necessitates a plant with extensive waste-sorting facilities to remove and recycle metals, glass, and where possible plastics. The remaining materials will be predominantly organic, and it is this organic material that is broken down in the pyrolysis reactor.

Before entering the reactor, the waste will be shredded into small particles. This will then be fed into a continuous grate system or loaded into a batch pyrolysis reactor. Pyrolysis may be carried out in the presence of steam, as in the Japanese example above, but this is not necessary and depends on the system design.

There are actually two products of pyrolysis of MSW in many pyrolysis plants, a combustible gas, and a char that is mostly carbon and can also be burned as a fuel. The pyrolysis process produces a gas with an energy content of up to 40 MJ/m^3, as high as natural gas, and this can be used in a variety of power-generation systems. This combustible gas is likely to contain some toxic compounds such as dioxins and furans. These are destroyed during the combustion of the syngas provided the combustion temperature is high enough.

Energy production in a pyrolysis waste-to-energy plant depends on how the plant operates. Some of the gases produced from the process must be used to generate heat to drive the pyrolysis reaction. This is most simply achieved by burning the products of pyrolysis in a conventional boiler to generate high-temperature steam that can be used to drive the process, with surplus steam used to drive a steam turbine. Other configurations are possible in which piston engines or gas turbines are the prime movers and drive generators to provide electricity, but steam-turbine generation systems will often be the most cost effective. The flues gases exiting the combustion process must be cleaned before release into the atmosphere, but as with gasification, the cleanup in normally simpler than for a mass-burn plant because fewer pollutants are produced.

In addition to the pyrolysis of MSW, there are a range of pyrolysis reactors that are designed to convert other wastes into useful products. These include pyrolysers that can handle vehicle tyres, plastic waste, wood waste, and other biomass wastes. Depending on the specific waste, these processes may be carried out at lower temperature than the pyrolysis of MSW so that they produce an oil product instead of a syngas. These products are typically called bio oil or tyre oil, depending on the waste being processed, and they can have a variety of uses.

Although gasification and pyrolysis are normally separate processes, there are configurations in which the two are combined. An example of this would be a pyrolysis plant treating separated solid waste to produce syngas and a solid-char residue. The residue is combustible and contains a high proportion of carbon so this can be gasified to produce more syngas and an ash or slag residue. The residue may be usable as a building material, while the additional syngas can be used for power generation.

PLASMA GASIFICATION OF WASTE

Plasma gasification is a very high-temperature process for converting waste into a fuel gas. The process uses extreme temperatures that can be in excess of $5000°C$. Under these conditions, the products of waste gasification are a relatively clean gas, a glass like solid slag from the residual ash, and in some cases molten metal.

A plasma is a special state of matter, sometimes called the fourth state, in which the atoms and molecules within it are ionized, creating a sea of positively and negatively charged particles. Any materials placed in the plasma will also dissociate into atoms or ions, with the precise dissociation products depending on the time they spend in the high temperature zone. These then reform as simpler molecules when the sea of particles cools. The technique has a variety of applications. In the case of waste treatment, it can be used to produce a syngas in a similar way to gasification or pyrolysis. The advantage for waste treatment is that all the potentially toxic compounds in the waste are destroyed.

The plasma that is used to heat the waste material is created by passing an extremely high voltage through a gas in a special chamber. The gas can be air, oxygen, nitrogen, or an inert gas such as argon. The choice will depend on the conditions required in the gasifier. The high voltage ionizes the gas between the electrodes of the plasma torch, allowing it to conduct electricity and a current flows through the ionized gas. This current generates a very high energy zone where the temperature can approach that of the sun. When waste material is passed through this high temperature zone in the plasma, it is immediately volatilized. The process can be carried out in the absence of air, in which case it is similar to pyrolysis, or it can be carried out with the addition of some air, or air and steam, in the same way as gasification. The products will be similar too. A cross-section of a plasma gasifier is shown in Fig. 7.3.

When the waste enters the high-temperature plasma, the waste materials are rapidly broken down from large and complex molecules

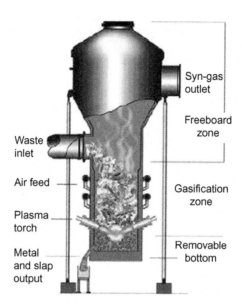

Figure 7.3 Cross-section of a plasma gasifier. Source: Westinghouse.[2]

into their constituent atoms. This will destroy any hazardous compounds that the waste may contain. The extremely high temperatures make the atoms and ions in the plasma very reactive and as they exit the plasma zone they will quickly reform into simpler molecules. The major products of the plasma gasification are hydrogen, carbon monoxide, and some hydrocarbons such as methane. The proportions of these will depend upon whether the process is carried out in the presence or absence of air. Although there may still be some larger molecules formed, the production of more complex molecules is reduced compared with other forms of waste incineration. In a Japanese demonstration plant commissioned in 1999 at Yoshii, the proportion of dioxins produced was 100 times lower than from a mass-burn incineration plant.

There are two different configurations that can be used for the plasma arc generation. The simplest is that described above in which two electrodes are used to create a discharge between them and this plasma discharge is used to process the waste that is passed through the plasma zone. This is called a nontransferred plasma. An alternative is to use a single electrode, with the second being supplied by the

[2]From a Westinghouse diagram published on NETL website

conductive lining of the plasma reactor vessel. This is called a transferred plasma system. The particular design can affect the performance of the reactor when processing waste. For example, with a nontransferred plasma arc, the heating of the waste may be nonuniform, leading to higher levels of toxic or pollutant products from the process. This may make the way in which the waste is prepared for the process more critical.

Creating the plasma is highly energy intensive. However, it appears to be relatively efficient. According to one manufacturer, only between 2% and 5% of the energy from the waste is needed to drive the plasma torch and all the remaining energy can be captured for further use[3].

Plasma gasification can be carried out using a range of reactors similar to those used for normal gasification, with updraft reactors common. The syngas from a plasma gasification plant can be used as an industrial feedstock, but it will more normally be burnt to generate electricity. The gas may be used to fire a gas turbine or gas piston engine, or the plasma gasification system can be combined with a conventional steam cycle power plant.

A demonstration project of this type in Israel used a fixed-bed updraft plasma gasification reactor with four plasma torches in the upper part of the reactor chamber. The temperature within this zone reached 6000°C. High temperature air and steam, at 1000°C, were injected into the lower part of the reactor. The reactor produced a combustible gas which was taken from the top of the reactor, while the inorganic material in the waste was turned into a molten slag that was collected from the bottom. From the plasma reactor, the product gas was taken directly into an afterburner where it was burnt in air and the hot flue gases were used to raise steam to drive a steam turbine, and to provide the power for the plasma gasifier. Emission cleanup included an acid—gas scrubber and a baghouse filter[4]. Depending on the gasification conditions, the efficiency of the gasification process was up to 60%.

---

[3]See the New Energy Technology Laboratory page on plasma gasification: https://www.netl.doe.gov/research/Coal/energy-systems/gasification/gasifipedia/westinghouse.
[4]Gasification of municipal solid waste in the Plasma Gasification Melting process, Qinglin Zhang, Liran Dor, Dikla Fenigshtein, Weihong Yang and Wlodzmierz Blasiak, Applied Energy, 2011.

CHAPTER 8

Waste to Energy Plants and the Environment

For most of the past century, urban waste has been considered a matter of management. As more waste has been produced, so more creative ways of dealing with it have had to be found. The incineration of municipal solid waste was one of the outcomes alongside the widespread use of landfill. Combustion of waste was effective in reducing the volume of waste in need of disposal and this was particularly important in island nations such as Japan with limited space for landfill. Where it was used, energy production was normally a secondary consideration.

Landfill, while cheap and relatively effective, has never been a particularly attractive solution to the problem of waste. Well-managed sites can keep the local impact to a minimum, but the daily transportation of waste and the noise, odors and general inconvenience caused by the process are often a burden on the local region. The combustion of waste is potentially much simpler and cleaner. However, the mass burn incineration of waste can lead to the release of a range of potent toxic materials including heavy metals, dioxins, and furans. This has caused public alarm in many parts of the world and has often made it difficult to build incineration plants.

Toward the end of the 20th century and into the 21st century, waste has become an environmental issue as well as a management issue. Waste is now considered a resource that must be used as fruitfully as possible. There is a waste hierarchy, as discussed in Chapter 2 which attempts to ensure the best use is made of any waste material. The combustion of solid waste is low in this hierarchy and combustion without energy capture is not considered viable today.

In modern developed societies, generating energy from waste should be one of the last resorts, in theory at least. In practice, waste burning in one form or another is widely practiced and is likely to continue to be a major means of managing urban waste well into the century. Arguments will continue about whether this represents the best use of

Energy from Waste. DOI: http://dx.doi.org/10.1016/B978-0-08-101042-6.00008-X

waste. Meanwhile, the practical concerns are to see that this is carried out safely.

The safety of waste to energy plants is a primary concern because the technology has earned a bad reputation in some parts of the world and there is significant public resistance to the construction of new plants. Persuading the public that waste to energy plants are safe is therefore a major issue within the industry.

Mass burn waste to energy incineration plants can produce a range of toxic or environmentally hazardous emissions that must be controlled. Advanced waste to energy technologies appears capable of reducing these emissions to a lower level than the older incinerators. In both the cases, there are two broad groups of products from the plants, gaseous emissions, and a solid ash. Both are potentially dangerous and must be carefully managed.

In addition, a waste to energy plant will bring a range of other impacts with it. These plants rely on a continuous supply of material to burn and this must be delivered regularly, usually from waste collection trucks. These will add to local traffic flows. The waste plant may be a sorting and recycling plant in addition to providing energy from waste. The sorting can give rise to a range of airborne pollutants if not managed correctly, and after waste has been sorted, some will be shipped out, while the residues from the waste to energy plant must also be removed to another site.

It should also be remembered that waste to energy plants of any type are combustion plants which burn waste, or a gaseous product of the waste, to generate electricity. This combustion produces carbon dioxide which is released into the atmosphere. There is therefore a global warming angle to waste incineration too which must not be overlooked.

There is one further environmental issue concerning waste to energy plants. As already noted, a large installation of this type requires a steady and secure supply of waste if it is to operate effectively and economically. Construction of such plants is expensive and once one has been built, economical operation may mean that the local region is tied to the continued combustion of waste in the plant. Some environmentalists argue that this can lead to combustion being preferred at the expense of more recycling and remanufacturing. This is a debate that is likely to continue for years if not for decades.

EMISSION LIMITS FOR WASTE COMBUSTION

The combustion of waste in waste to energy plants must meet strict environmental standards in most parts of the world for emissions into the atmosphere. These standards cover the amount of a range of materials that can be released in each unit volume, usually a cubic meter,[1] of flue gas. The actual regulatory amounts will vary from jurisdiction to jurisdiction, but the problem remains the same.

The composition of municipal waste is always uncertain. It may contain toxic metals that will be volatilized if heated and plastics which can be converted into dangerous chemicals such as dioxins. Using it as a combustion fuel means that all these elements and compounds, and many others, can be released into the atmosphere.

Current regulations have been formulated partly in response to historical emissions from waste combustion plants. For example, in the late 1980s the US Environmental Protection Agency (EPA) listed waste to energy plants as a major source of mercury and dioxins in the environment. To counter this, "maximum available technology" regulations were introduced into the USA in 1995. EPA data showed that in response, toxic equivalent (TEQ) emissions from the majority of the United States waste to energy plants fell from 4260 g in 1990 to 12 g in 2000.

Table 8.1 shows limits for a range of emissions for plants operating in the European Union. The figures in the table are daily average emission concentration limits. Actual emissions are expected to fall below these values. However, shorter term levels can be higher. For example, 30 minutes averages can be two or three times higher than the daily-average maximum.

Dust, which covers a range of particulates that can be produced during combustion, must be below $10 \, mg/m^3$. This is lower than the limit for a coal-fired power station that can emit up to $20 \, mg/m^3$. The limit for volatile organic compounds that includes a range of potential molecular pollutants, calculated as the total amount of organic carbon they contain, is also $10 \, mg/m^3$. Acid gas emissions are strictly controlled. For hydrogen chloride, the limit is $10 \, mg/m^3$, and for hydrogen fluoride it is $1 \, mg/m^3$. The amount of sulfur dioxide that can be

---

[1]The figure is often expressed as the amount in each Normal cubic meter, which is a standard volume of gas at specified temperature and pressure.

Table 8.1 EU Daily Average Air Emission Limits

Pollutant	Emission Limit (mg/m^3)
Total dust	10
Volatile organic compounds	10
Hydrogen chloride	10
Hydrogen fluoride	1
Sulfur dioxide	50
Nitrogen oxides for plants >6 t/h	200
Nitric oxide for plants <6 t/h	400
Carbon monoxide	50

Source: UK Government.[2]

Table 8.2 Average Limit for Heavy Metal Emissions

Metal	Emission Limit (mg/m^3)
Cadmium	0.05
Thallium	0.05
Mercury	0.05
Antimony	0.5
Arsenic	0.5
Lead	0.5
Chromium	0.5
Cobalt	0.5
Copper	0.5
Manganese	0.5
Nickel	0.5
Vanadium	0.5

Source: UK Government.[2]

released is 50 mg/m^3, while for nitrogen oxides, the limit depends on plant size. For plants that can process more than 6 t/hour, the limit is 200 mg/m^3. For smaller plants with capacities of 6 t/hour or lower, the limit is 400 mg/m^3. Carbon monoxide emissions are also controlled with a limit of 50 mg/m^3.

Heavy metals are a major concern too. Table 8.2 lists the metals that might be released and the limits set on their emissions. Those considered the most toxic are cadmium, thallium, and mercury. The limits

[2]Environmental Permitting Guidance: The Waste Incineration Directive, UK Government, 2010.

shown in the table are for a sample period of between 30 minutes and 8 hours. For these three, the maximum allowable is 0.05 mg/m^3. The other metals listed in the table are antimony, arsenic, lead, chromium, cobalt, copper, and manganese. For all these, the limit imposed is 0.5 mg/m^3.

Dioxins are a range of chemically related environmental pollutants that persist in the environment. According to the World Health Organization (WHO), they are found throughout the world and they accumulate in the food chain, primarily in the fatty tissue of animals. Humans are exposed because they eat animals that contain dioxins. The WHO says "dioxins are highly toxic and can cause reproductive and developmental problems, damage the immune system, interfere with hormones and also cause cancer." The toxicity of dioxins and of other related compounds is measured by a factor called the TEQ. This relates the toxicity of each compound to a reference compound, 2,3,7,8-tetrachlorodibenzodioxin, the toxicity of which is set at 1. The original TEQ scheme was set up by North Atlantic Treaty Organization (NATO) in 1989 and is often referred to as I-TEQ. The WHO has recently suggested a modified scheme, WHO-TEQ. The reference limit of emissions of the total TEQ of such compounds from a waste to energy plant is 0.1 ng I-TEQ/m^3.

There are other compounds that should be monitored too. These include a range of polycyclic aromatic hydrocarbons. In addition, the composition of waste water from a plant should be monitored for toxic content.

ENVIRONMENTAL POLITICS AND SCIENCE

Whatever be the science of waste combustion for electricity generation, there is and will continue to be a strong environmental lobby that opposes all waste incineration. In some cases, the opposition reflects fear of the effects of a waste plant, particularly of its emissions; in others it is based on arguments about the best way to use waste and whether waste incineration is a sensible option. The waste to energy industry is generally alive to these concerns and where possible attempts to allay them.

As far as airborne emissions are concerned, current scientific observations suggest that modern waste to energy plants contribute little to the level of prescribed pollutants found in the environment. It is true, nevertheless, that they do produce some emissions. The level may be much less than the overall background level of the pollutants in the environment but objections based on the aim to eliminate all pollutants cannot be countered. Then it becomes a matter for society to decide, which is where the politics enter.

The question, whether the combustion of waste should ever take place in the best of all worlds, is more difficult to answer. In principle, virtually all waste could be recycled with only a tiny residual proportion remaining that probably needs to be buried. With careful industrial planning for reuse, even that might be avoided. In a waste system where extensive sorting and recycling already takes place; the waste that ends up at a combustion plant is normally primarily organic waste. This could be turned into compost and then returned to the soil. Today, this is often impractical and that is why it is burned. There may come a time when the reuse of even this organic material becomes normal. If that happens, then most waste to energy plants will disappear. Today, that appears a long way into the future.

CHAPTER 9

The Economics of Energy From Waste

Waste-to-energy combustion plants are complex industrial installations and they are expensive to build. Since the capital cost weighs heavily on the economics of a power station, electricity produced by these plants is not competitive with other sources of electricity when treated solely as an electricity generation station. Some other types of waste-to-energy plant, such as landfill gas or anaerobic digestion units, can offer more cost-effective sources of electricity, while waste-to-energy plants that burn agricultural biomass wastes can also be economical to run. These latter are normally considered to be biomass plants. Municipal solid waste may also be considered a renewable source in some jurisdictions.

Fortunately for plants that burn municipal solid waste to generate power, the economics do not rely solely on the sale of electricity. The disposal of municipal waste is an expensive business no matter how it is handled and there are fees associated with disposing of municipal waste. A waste-to-energy combustion plant can expect to earn an income from the waste it accepts, often known as a gate fee or tipping fee, and this will greatly affect the overall economics. Besides that, environmental legislation in some regions now encourages that all options be pursued before waste is buried in a landfill site, and this encourages the use of combustion as a preferred alternative to landfill.

The economics of a waste-to-energy combustion plant will depend upon how the tipping fee compares to the cost for burying the same waste in a landfill site. Table 9.1 compares the average tipping fees for waste-to-energy plants and landfill sites in three different countries, Sweden, the United Kingdom, and the USA. In Sweden, the tipping fee for a landfill site is $193/t of waste, while for a waste-to-energy plant the fee is $84/t. The very large difference encourages the use of waste-to-energy plants since the economic advantages are significant. In the United Kingdom, the average tipping fee for a landfill site is $153/t, while for a waste-to-energy plant, the average fee is $148/t.

Energy from Waste. DOI: http://dx.doi.org/10.1016/B978-0-08-101042-6.00009-1

Table 9.1 Tipping Fees for Waste to Energy and Landfill		
Country	Average Tipping Fee for a Waste-to-Energy Plant (US$/t)	Average Tipping Fee for a Landfill Site (US$/t)
Sweden	84	193
United Kingdom	148	153
USA	68	44
Source: World Energy Council.[1]		

The difference here is much smaller than in Sweden but can still help tip the balance in favor of waste combustion. In the USA, in contrast, the tipping fees for landfill sites average $44/t, while the fee for waste-to-energy plants averages $68/t. Not surprisingly, the use of waste-to-energy plants in the USA is much lower than in Sweden.

WASTE-TO-ENERGY PLANT COSTS

A study carried out for the Mayor of London and published in 2008 looked at the cost of the principle waste combustion technologies. The main findings are shown in Table 9.2. The study concluded that a conventional incineration facility would cost around £45 m for a plant with the capacity to treat 100,000 t/year of municipal solid waste (MSW), while for a 200,000 t/year plant, the cost would be £76 m. With the maximum power output from the smaller plant put at 6 MW, this equates to a capital cost of £7500/kW, while the larger plant has a maximum output of 12 MW, equating to a capital cost of 6300/kW.

Advanced thermal treatment plants such as gasifiers and pyrolysis plants have slightly higher costs, as shown in the table. Their potential power outputs are also slightly lower. As a consequence, the capital cost of a 100,000 t/year advanced plant is £9100/kW, while for the 200,000 t/year plant, the capital cost is £7700/kW. Operating costs for the plants are broadly similar at between £40/t and £70/t of capacity depending upon plant size.

These UK costs are similar to estimates for plant costs in the USA where the cost of a typical municipal waste combustion plant was put at $5000/kW to $10,000/kW during the middle of the first decade of the 21st century. Again, smaller plants are relatively more expensive than larger plants.

[1]World Energy Council, World Energy Resources: Waste to Energy 2016.

Table 9.2 The Cost of Waste-to-Energy Plants in the United Kingdom		
Plant Waste Treatment Capacity (kt/year)	**Conventional Incineration (m)**	**Advanced Thermal Treatment (m)**
100–115	£45	£50
150	£60	£68
170–200	£76	£85
Source: Mayor of London.[2]		

To put these costs into perspective, figures from the US Energy Information Administration for its 2016 Annual Energy Outlook show that the cost of a new modern coal-fired power plant in 2015 was $4649/kW, and for a new nuclear plant, it was $5288/kW. From the same source, the cost of a new onshore wind farm was $1536/kW, and for a solar photovoltaic power plant, it was $2362/KW. Clearly, a waste-to-energy combustion plant would not be an economical choice if the sole aim was to produce electricity as cheaply as possible.

LEVELIZED COST OF ELECTRICITY FROM WASTE-TO-ENERGY PLANTS

The cost of electricity from a power plant of any type depends on a range of factors. First, there is the cost of building the power station and buying all the components needed for its construction. In addition, most large power projects today are financed using loans so there will also be a cost associated with paying back the loan, with interest. Then, there is the cost of operating and maintaining the plant over its lifetime, including fuel costs if the plant burns a fuel. (In the case of the waste-to-energy plant, this is an income rather than a cost.) Finally, the overall cost equation should include the cost of decommissioning the power station once it is removed from service.

It would be possible to add up all these cost elements to provide a total cost of building and running the power station over its lifetime, including the cost of decommissioning, and then dividing this total by the total number of units of electricity that the power station actually produced over its lifetime. The result would be the real lifetime cost of electricity from the plant. Unfortunately, such calculation could only be completed once the power station was no longer in service. From a practical point of view, this would not be of much use. The point in

[2]Costs of incineration and nonincineration energy-from-waste technologies, The Mayor of London, 2008.

time at which the cost-of-electricity calculation of this type is most needed is before the power station is built. This is when a decision is made to build a particular type of power plant based normally on the technology that will offer the least cost electricity over its lifetime.

In order to get around this problem, economists have devised a model that provides an estimate of the lifetime cost of electricity before the station is built. Of course, since the plant does not yet exist, the model requires that a large number of assumptions be made. In order to make this model as useful as possible, all future costs are also converted to the equivalent cost today by using a parameter known as the discount rate. The discount rate is almost the same as the interest rate and relates to the way in which the value of one unit of currency falls (most usually, but it could rise) in the future. This allows, for example, the cost of replacement of a plant component 20 years into the future to be converted into an equivalent cost today. The discount rate can also be applied the cost of electricity from the power plant in 20-year time.

The economic model is called the levelized cost of electricity (LCOE) model. It contains a lot of assumptions and flaws but it is the most commonly used method available for estimating the cost of electricity from a new power plant. One particular problem is that the model does not take into account cost risks. For example, the cost of natural gas can fluctuate widely so that it may be cheap to buy gas when a plant is built, but 5 years later the cost is so high that operation of the plant is uneconomical. The level at which the discount rate is set can also be problematical. It is typical to use a discount rate of 5% and 10% in calculations. However, in the middle of the second decade of the 21st century, the actual interest rate is close to zero.

The LCOE model can be applied to any type of power station, including a waste-to-energy plant. Table 9.3 shows figures for the

Table 9.3 Levelized Cost of Electricity from Waste-to-Energy Plants	
Technology	US Levelized Cost ($/MW h)
Waste-to-energy combustion plant	80–210
Landfill gas	45–95
Source: World Energy Council/Bloomberg.[3]	

[3]World Energy Perspective: Cost of Energy Technologies, World Energy Council/Bloomberg New Energy Finance, 2013.

LCOE for two types of waste-to-energy plant. The estimated cost for electricity from MSW combustion, waste-to-energy plant in the USA is $80–210/MW h, while for a landfill gas installation in the USA, the LCOE is $45–95/MW h. The estimates for similar plants in Western Europe are the same. Landfill gas installations in China can produce electricity for $34–83/MW h, slightly lower than in the USA or Europe.

Another estimate of the costs in the USA put the cost of energy from a landfill gas installation in 2007 at around $82–99/MW h, while for anaerobic digestion, the cost of electricity was $62–128/MW h.[4] Landfill gas is usually a cost-effective way of generating power because the fuel is free. Anaerobic digestion of waste is relatively more expensive but can still provide cost-effective power when supplying energy directly to end users where the competition is with the retail cost of the same electricity from the grid. Anaerobic digestion plants and landfill gas plants are normally small-scale installations that operate in this way.

[4]Arizona Renewable Energy Assessment, Arizona Public Service Company, Salt River Project, Tucson Electric Power Corp, prepared by Black and Veatch Corp, 2007.

INDEX

Note: Page numbers followed by "*f*" and "*t*" refer to figures and tables, respectively.